TRIZ

FOR DUMMIES®

A Wiley Brand

TRIZ
FOR DUMMIES®
A Wiley Brand

by Lilly Haines-Gadd

FOR DUMMIES®
A Wiley Brand

TRIZ For Dummies®

Published by: **John Wiley & Sons, Ltd., The Atrium, Southern Gate, Chichester,** www.wiley.com

This edition first published 2016

© 2016 by John Wiley & Sons, Ltd., Chichester, West Sussex

Registered Office

John Wiley & Sons, Ltd., The Atrium, Southern Gate, Chichester, West Sussex, PO19 8SQ, United Kingdom

For details of our global editorial offices, for customer services and for information about how to apply for permission to reuse the copyright material in this book, please see our website at www.wiley.com.

For general information on our other products and services, please contact our Customer Care Department within the U.S. at 877-762-2974, outside the U.S. at 317-572-3993, or fax 317-572-4002. For technical support, please visit www.wiley.com/techsupport.

Wiley publishes in a variety of print and electronic formats and by print-on-demand. Some material included with standard print versions of this book may not be included in e-books or in print-on-demand. If this book refers to media such as a CD or DVD that is not included in the version you purchased, you may download this material at http://booksupport.wiley.com. For more information about Wiley products, visit www.wiley.com.

A catalogue record for this book is available from the British Library.

Library of Congress Control Number: 2016931978

ISBN 978-1-119-10747-7 (pbk); ISBN 978-1-119-10748-4 (ebk); ISBN 978-1-119-10749-1 (ebk)

Printed and Bound in Great Britain by TJ International, Padstow, Cornwall.

10 9 8 7 6 5 4 3 2 1

MIX
Paper from
responsible sources
FSC® C013056

Contents at a Glance

Table of Contents

Introduction

· ·

*A*re you looking for a better way to solve problems or interested in developing your creative ability? Or are you seeking a method for generating innovative thinking, in yourself and in the team around you? Perhaps your organisation has become so lean and efficient that you don't know how to get people thinking innovatively again. Or maybe you want to find a way to encourage everyone to share and develop ideas together as a team. If any of these ring true, or you're just looking for a novel new way of navigating the many rapids you encounter along the river of life, TRIZ will help you. You've come to the right book!

TRIZ is the outcome of extensive research into patents and scientific journals. It has a wonderful engineering and scientific pedigree, and as a result, many case studies and examples you'll hear about are technical.

However, and this is the important thing, it works – and I've seen it work – on any kind of problem: management problems, business problems, social problems, even personal problems. What TRIZ gives you is the ability to think very clearly and creatively. You unpick thorny issues, define your problem correctly and are then given useful suggestions regarding how these problems may be solved. You use your brain to the best of its ability and, in fact, many of the TRIZ tools are based around changing the way you think by repeating the thinking patterns of the most creative and successful problem solvers. These principles are helpful regardless of the subject matter.

TRIZ gives you both skills in creative thinking and confidence in your creative ability. After reading this book you'll be able to put some TRIZ magic into practice and enhance and extend your natural creativity and problem-solving ability.

TRIZ is a Russian acronym that stands for 'Teoriya Resheniya Izobretatelskikh Zadatch', which translated into English means 'Theory of Inventive Problem Solving'. As this is a bit of a mouthful, we generally use the acronym, but it's helpful to know what it stands for; in fact, TRIZ is more than just a theory, it's a practical toolkit, a method, a set of processes and even a bit of a philosophy to help you understand and solve problems in clever ways.

Genrich Altshuller, the mastermind behind TRIZ and a very clever engineer, scientist, inventor and writer (of both TRIZ books and science fiction), asserted that 'anyone can be creative'. TRIZ shows you how!

About This Book

This book isn't the ultimate guide to TRIZ. Instead, it gives you a short (and hopefully fun) introduction to the TRIZ tools and processes, based on how it can be used (and how I use it at work), rather than explaining every detail of the theory. You can start anywhere because each chapter stands alone and provides insights into new ways of thinking about problems and finding solutions. As you read you'll come across occasional sidebars, which contain additional information about the topics at hand (and, if you're lucky, the occasional joke). Reading these is optional because they're not necessary to understanding the main text.

Most books about TRIZ are written from a technical starting point. I read many when I first encountered TRIZ but found them hard to read, because most of the examples were based on engineering problems that I struggled to understand. Even the non-technical examples were often described in language that seemed very technical to me (most TRIZ experts are engineers) and didn't reflect how simple the tools can be to pick up and use – for anyone. So in this book, I make things as straightforward as I can.

As TRIZ has always been used for any kind of problem solving, I thought it was time for a general book that even a dummy like me would understand. To that end, every example is general and non-technical. Even when I describe a technical system, I use everyday items that you've probably used, such as a cheese grater or toilet brush (not together!). I'm regularly asked, 'Will TRIZ work for me even though I'm not an engineer?' I wrote this book to (hopefully) show you that the answer is *yes*! And to start to show you how.

Because I describe things as simply as possible, I may use slightly different terminology to what you might find elsewhere. TRIZ was first conceived in Russian and then translated into other languages, not always consistently. On top of that, it's developed by a whole community, not one individual, and isn't owned by any one person. As a result, a number of different terms are used for the same tools, and some tools are arranged in different ways. Some derivative approaches also exist, whereby people have tried to extend or develop TRIZ and created their own toolkits and methods as a result. In this book, I describe 'classical TRIZ', which is based only on the framework developed in the original research.

The terminology in this book is based on the Oxford TRIZ approach: classical TRIZ described simply (with nothing added or taken away).

Foolish Assumptions

I make only one assumption about you: that you're interested in learning something new. This is the first step to thinking creatively and becoming a better problem solver – and you've demonstrated it very effectively by picking up this book!

Your expertise or discipline is irrelevant; however, some of the technically based tools are easier to grasp if you have some technical knowledge or at least an understanding of how the physical world works, that is, an interest in science. That said, the examples I use are broad enough that anyone will get them – I wrote the book for a dummy like me, after all!

The only other useful attribute is that you've encountered problems, but then, who hasn't?

Icons Used in This Book

To help you pick out the information most useful to you, I've used a few graphical icons in the book to highlight key details. Whenever you see the following icons in the margin, this is what you can expect from that paragraph:

A TRIZ tip or trick for improving your thinking or developing your problem-solving skills.

This icon reminds you of something that you should bear in mind when applying a specific bit of TRIZ thinking.

This icon warns you of common mistakes or pitfalls that could trip you up when you're applying TRIZ to a problem.

Throughout the book, I've included helpful real-life examples. Many other TRIZ books focus on engineering problems, but I've applied my TRIZ examples in a broader context.

Beyond the Book

In addition to the material in the print or e-book you're reading right now, this product also comes with some access-anywhere goodies on the web. Check out these features:

- ✔ **Cheat Sheet** (www.dummies.com/cheatsheet/triz): A handy little guide to refer to as you read through this book.
- ✔ **Dummies.com articles** (www.dummies.com/extras/triz): You can also find relevant online articles that supplement each part in this book with additional tips and techniques.

Where to Go from Here

If you're not sure where to begin and don't fancy the usual practice of 'Start at the beginning', here are my top suggestions:

- ✔ The 40 Inventive Principles are the most famous TRIZ tool, and are used to solve contradictions. Chapter 3 is a good place to start as it gives you not only an overview of this popular and powerful tool but also some background into the logic of TRIZ.
- ✔ If you work on developing new products, Chapter 4, which covers the Trends of Technical Evolution, shows you the likely directions your systems will take in the future.
- ✔ Chapter 7 offers a good introduction to the psychological blocks to creativity and how to get over them, using the suite of TRIZ creativity tools.
- ✔ Chapter 8 introduces Thinking in Time and Scale, a deceptively simple tool for stretching your thinking, restructuring your view of a problem and generating innovative solutions.

Having said that, you can start at the very beginning because, as the fresh-faced trainee nun in *The Sound of Music* says, 'it's a very good place to start'!

Part I
Getting Started with TRIZ

getting started
with
TRIZ

In this part . . .

✔ Get an introduction to the TRIZ tools, process and fundamental logic of innovative problem solving.

✔ Understand the TRIZ philosophy and learn how to start thinking like a genius.

Chapter 1

Going from Zero to TRIZ

*W*e've all got problems, right? And largely we can work out how to solve them, even when the problems seem really tough. Human beings are designed to be problem solvers, and we're generally really good at it, so why do we need to go back to the drawing board and learn a new way to tackle problems?

Well, because it's possible to learn from each other – and from problem solvers in the past. TRIZ is an attempt to try to cut across different disciplines and 'bottle' the fundamental logic of problem solving for everyone no matter what their job, speciality or area of expertise.

The greatest achievements in the arts and sciences have come about because people have been able to build on the previous work of others. When developments and breakthroughs have occurred – whether the drawing of perspective in art or the theory of gravity or the discovery of DNA – they've been shared so they can be built upon rather than rediscovered over and over again. However, these developments, and the preceding problems and solutions, are typically described in the language of the discipline in which they happened. As a result, only people with specialist knowledge are truly capable of understanding these developments. While this situation's great for them, it cuts out everyone else. Because problem solving is seen as being specific for each discipline – the assumption being that lawyers, for example, must face very different problems to chemists – people tend to stay within their own industry and field of expertise when they face problems and are looking for solutions.

TRIZ takes the opposite approach.

One of the cornerstones of TRIZ is that the same problems occur again and again across different disciplines and applications, and that people are constantly reinventing the wheel by solving them from scratch every time. At the heart of TRIZ is the belief that, if you can understand how your problem is similar to someone else's, you can reapply his clever solutions.

When you use TRIZ, you're able to access the clever thinking of genius problem solvers from all areas of science, engineering and technology and can reapply what they've learned. You don't reinvent the wheel – you find new and exciting ways of and ideas for using clever existing concepts to give you what you want.

And generating new ideas *will* be very easy for you because you have TRIZ. If you need solutions to a problem, you can just apply a simple thinking tool. If you hit a dead end, hit the problem with TRIZ. If you have a solution that looks pretty good, improve it even more by teasing out its problems and solving them. You can always do more TRIZ, which means that solutions and improvements are always out there to be discovered. It's an exciting journey, and you and the people you're making it with will appear to be geniuses as you find the right solutions to all the problems you encounter along the way.

Getting to Know TRIZ

TRIZ subdues complexity and keeps detail in its place. TRIZ logic demands that you have a clear idea of where you are and where you're going, which helps you keep your eye on the prize and avoid getting tripped up with irrelevant detail, waylaid by trivial issues or seduced by premature solutions.

Increasing Ideality

The main goal of TRIZ is to increase Ideality. *Ideality* is the TRIZ equation for working out how good something is, as shown in Figure 1-1.

The Ideality of a system is the ratio of its benefits compared to its costs and harms:

- ✔ **Benefits** are all the outputs that you want, expressed as outcomes (not solutions).
- ✔ **Costs** are all the inputs required to create a system (not just money but also time, materials, cleverness and so on).
- ✔ **Harms** are all the outputs from your system that you don't want (even neutral things that aren't actively harmful).

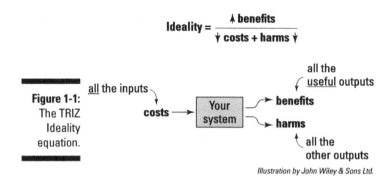

$$\text{Ideality} = \frac{\uparrow \text{ benefits}}{\downarrow \text{ costs + harms } \downarrow}$$

Figure 1-1:
The TRIZ
Ideality
equation.

Illustration by John Wiley & Sons Ltd.

A *system* in TRIZ is a very general term: it means any kind of product or process that's created and used to meet a need.

Ideality is important because it's very simple, and very brutal. It holds in the front of your mind the reason you're doing whatever it is you're doing. The benefits are the outcomes that you want but no mention is made of how you get those benefits. That's deliberate because it keeps your focus on the outcomes you want and not on exactly how you'll achieve them. This approach stops you becoming enraptured with solutions too soon, and always reminds you that other ways of getting what you want may exist. When you think about benefits, you consider all the things you want and not merely the outcomes you believe are achievable. This drives you continually to find new benefits you can deliver, and ways to increase the levels of benefits you're currently achieving.

You're also aware of *all* the downsides associated with the various ways of getting what you want. This is important because it forces you to look for problems, which means in turn that you'll be able to solve the problems and improve your system continually, in an iterative way.

Ideality identifies two kinds of problems:

- ✔ Costs (all inputs)
- ✔ Harms (all outputs you don't want)

TRIZ is always looking for ways to reduce *costs*; not just money but also time, parts, materials, effort – any kind of input required to create your system, in fact. TRIZ thinking pushes you towards creating simple, elegant systems and solutions to problems, which often involves finding innovative ways of getting what you want. While many traditional approaches also consider both costs and benefits (or sometimes functions), thinking about harms provides additional power.

Harms are all outputs you don't want – they needn't be actively harmful but are things produced by your system that aren't useful to you. Examples include things that may seem 'neutral' initially, such as heat from a laptop or noise from a washing machine, any complicated features you don't use on a smartphone, and waste or even potential risk. Thinking about harms encourages a more holistic view of your system, in which you consider its impact in the bigger picture. It also drives you towards simpler, more efficient systems, because all harms are things you're fundamentally paying for in some way: heat from a lightbulb may not be actively harmful but it is wasted energy, and finding a way to reduce that heat output will result in either more light (increased benefit) or reduced energy use (reduced cost).

All TRIZ tools exist to improve Ideality. They increase benefits, reduce costs or reduce harms – or all three! Ideality is referred to throughout this book because, while you can use it as a standalone tool (see Chapters 5 and 9 for details), it's also more of a fundamental way of understanding TRIZ and its purpose.

Ideality expresses in a nutshell the duality of TRIZ. On the one hand, you have one eye on utopia and all the benefits you want (even though you know you probably won't get them). On the other hand, you're searching for all the problems that exist in your real-world system (so you can get rid of them). TRIZ helps you connect fantasy and reality: you allow yourself to imagine perfection *and* engage with the nitty-gritty of practical systems. Obviously, this behaviour is a contradiction; however, TRIZ says the world is full of contradictions and you shouldn't be afraid of them, ignore them in the hope that they'll go away or compromise too soon in an attempt to resolve them. Ideality is a concept that balances the good and bad in any kind of system, and holds them together at the same time. Understanding and appreciating the conflict between the good and the bad allows you to work in an ambiguous, creative and potentially very fruitful space.

Uncovering patterns in human creativity

The logic underpinning TRIZ is that patterns exist across problems and the solutions that have previously been found to those problems. If you can understand how your situation is similar to previous situations, you can short-circuit the problem-solving process and generate very creative solutions.

TRIZ was observed, not invented. The earliest research found that the same problems occur again and again across different industries, and that very similar solutions are found to these problems (Chapter 2 gives you the lowdown on how TRIZ was developed).

For any problem you encounter, chances are that someone else will have seen something similar in the past – and found a solution. Even more excitingly, the solutions people come up with also exhibit similarities. What the TRIZ community has captured are the patterns that exist in both the kinds of problems that people address and the way in which they solve them. These patterns have been encapsulated in a series of thinking tools that the rest of us can apply to solve our problems.

Learning to think in the abstract

All TRIZ problem-solving tools help you move between thinking about very specific, real-world problems and considering more general, conceptual ways of looking at those problems.

You can view this process as a journey whereby, rather than attempting to go from where you are now directly to where you want to get to, you take a step out of reality into an abstract world. You then understand your problem in a more conceptual way and can create a generalised 'model' of it that identifies its true nature. When you've done this, you can look for abstract, generalised solutions to your problem, and then work out how to turn these abstract solutions into real, practical solutions. Lots of creativity tools exist to help you model conceptual solutions, but TRIZ is unique in providing lists of conceptual solutions based on previous successful innovations that you can apply at this point to find the right solutions to your problem (see the nearby sidebar, 'The four solution tools: Listy loveliness or 100 answers to everything'). After you've modelled your problem in a conceptual way, you're directed to a small number of conceptual solutions that will be useful for that type of problem. This process may seem a bit long-winded, but I promise it isn't! The time you spend grappling with your problem and modelling it in a conceptual way aids your clarity of thought and understanding and ensures the real problem is explicit. Looking up the solutions is easy, and only takes a few minutes. The time spent generating solutions is then enormous fun: you're being creative and thinking of answers to your hardest questions and problems but are also focusing all that brainpower and creativity in the most useful places – where you're most likely to find inventive and creative solutions.

A number of TRIZ tools help you take a specific problem and create a conceptual model of it. When you've created that model, you can then look up how people have solved this kind of problem in the past. A number of ways of solving this problem that other people have used successfully in the past will exist. You can then take these situations and reapply them to your situation.

Part of the power of TRIZ thinking and the TRIZ tools comes from this moving between the real world and the conceptual, more abstract world. This process is called using the *Prism of TRIZ*, as shown in Figure 1-2.

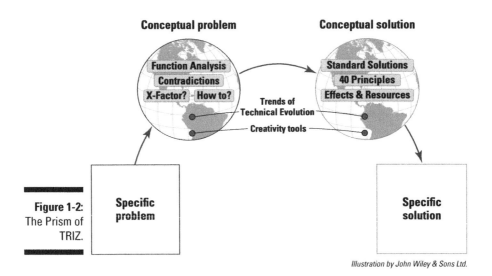

Conceptual problem

Function Analysis
Contradictions
X-Factor? — How to?

Conceptual solution

Standard Solutions
40 Principles
Effects & Resources

Trends of
Technical Evolution

Creativity tools

Figure 1-2:
The Prism of
TRIZ.

Specific
problem

Specific
solution

Illustration by John Wiley & Sons Ltd.

The four solution tools: Listy loveliness or 100 answers to everything

The completely unique aspect of TRIZ is the lists of solutions derived from patent databases: the 40 Inventive Principles, the 8 Trends of Technical Evolution, the 76 Standard Solutions and the Effects Database. These lists comprise the most clever and inventive solutions discovered in patents and scientific journals. Specific solutions to specific problems were distilled into the essence of what made them clever; solutions such as vibrate something, make it more flexible, insulate from something harmful. As well as generalising the solutions found in patents, the TRIZ community also described the problems it was solving in a conceptual rather than technical or scientific way, such as 'something gets better and something else gets worse' (a Technical Contradiction) or 'this is useful but not quite good enough' (an Insufficient Useful Action). A huge amount of work was then carried out cross-referencing the kinds of problems that re-occurred and the solutions that were most commonly used to solve them.

When you have a specific problem, you first convert it to a general problem. You then look up the general solutions to your problem in one of the four solution tool lists. You're directed to just a handful of conceptual solutions to your problem, which you have to convert into something real, practical and tangible in your specific situation – stepping through the Prism of TRIZ (see Figure 1-2). If finding solutions is like digging for buried treasure, TRIZ supplies the map. It shows you where to dig first: the places in which you're most likely to find the most clever solutions.

The TRIZ lists contain approximately 100 solutions. The mathematically minded among you will notice that this figure is less than the number you'd get if you added up all the solutions in the lists above – and that's because the lists of tools overlap. As a result, you don't need to use all the tools when looking for solutions: applying two or three will give you good coverage of the solution space.

Connecting conceptual thinking with your knowledge and experience

Using your experience and knowledge is a critical element of TRIZ. It's not just a question of taking someone else's solution and applying it directly; rather, you're given a conceptual prompt or trigger.

You then have to activate all your domain knowledge and experience of the problem and the situation in order to turn that conceptual solution into something real. The conceptual solutions need to connect with your practical expertise in order to become useful. As a result, TRIZ makes the best use of your experience – it is not a substitute for it.

The TRIZ problem-solving process utilises your knowledge, practical experience and expertise to the best of their ability. The TRIZ solution tools focus and enhance your experience, so that you use your knowledge in new and inventive ways. If you have no knowledge or experience in a particular area, you won't be able to solve problems in that area with TRIZ because you don't know how things work.

Here's something to bear in mind, as it applies to you as well as me: for all my TRIZ superpowers, I can't solve my clients' problems for them, as I don't have their domain knowledge. I can only help them understand and define their problems with TRIZ and look up the suggested solutions. That last step in the Prism of TRIZ – the leap from conceptual to practical solutions – is completely up to them. TRIZ stimulates their creativity and, as a result, they're able to generate insight, new thinking and many innovative solutions.

What's so heartening about this is that because TRIZ shows you how to apply your knowledge in new ways, you'll make better use of TRIZ and become a better problem solver as your career progresses. When you develop very deep expertise in an area, it can become like a narrow pit in which your thinking is stuck: you know solutions to many, many problems and you can think of them easily. So easily, in fact, that thinking of anything new is difficult. TRIZ helps you generate those solutions based on your experience, and then move beyond them to apply your expertise in novel ways. For those of us who aren't getting any younger (which, let's face it, is all of us), this is good news. It means that, once you've learned TRIZ, as your experience and expertise grow so will your creativity and problem-solving ability.

TRIZ enhances and develops expertise

While TRIZ can't replace expertise, it may well help it develop. One interesting observation from TRIZ workshops is that when teams of different people are working on a problem, those who have experience in the same field but without deep knowledge of the problem or particular area will often generate solutions that are both highly innovative and highly practical. When you're looking for solutions to a problem, therefore, don't just involve the experts. Instead, also involve people who have general domain knowledge in your area, even if they don't have specific knowledge of the problem at hand. TRIZ will help them generate useful solutions. This also suggests that if you're just starting out in your career, TRIZ will help you generate the same kind of solutions as experts in your field. You'll also have the benefit of developing flexibility of thinking as you gain expertise in your area.

Going beyond your own experience

One of the interesting things about different disciplines is that they often take different approaches to identifying problems and coming up with new ideas. If you put a teacher, a doctor, an engineer, a physicist, a mathematician and a philosopher together to solve a problem, they'll all have very different ideas about how best to examine it and find solutions (many jokes are based on this premise, and you can find one at the end of Chapter 2!). Your profession influences how you look at problems and the kind of solutions you generate.

If you want to improve the behaviour of a naughty child, how you characterise the problem and generate solutions depends on your perspective. Consider the perspectives of a parent (who needs to get everyone to school on time *and* teach the child patterns of behaviour for the longterm), a teacher (who may need to manage a whole classroom *and* get children to learn), a child psychologist (who may focus on the underlying causes of the naughty behaviour) and an anthropologist (who may be interested in how children and parents interact and communicate and what this says about the local culture). None of these approaches is wrong. Each has something good about it and will bring a different perspective to the problem that's new and interesting. However, everyone thinks their approach is 'the right one', and will tackle problems according to the kind of solutions they're familiar with (as the nearby sidebar, 'Tackling the glass of water problem', suggests, when you have a hammer, everything can look like a nail).

Tackling the glass of water problem

Here's an ice breaker to demonstrate different approaches to problem solving: put a glass of water on a table and ask people to remove the water without moving either the glass or the table.

Different people generally come up with different solutions. Mechanical engineers do things to the glass (for example, break it, drill a hole in the glass and the table, vibrate the glass); chemists do things to the water (for example, apply hydrolysis so that the water is divided into hydrogen and oxygen, change the water chemically into something else); biologists use living things (for example, have a person suck it up with a straw, put flowers in the glass to draw up the water). I encountered a parent who suggested leaving a toddler alone in the room for two minutes. 'You'll come back in, 'she asserted,' and the water will be all over the floor and the child will say, "I didn't touch it!"'

This problem can be tackled in lots of ways. Applying TRIZ to solving it will help you capture the knowledge of all the people in the team and go beyond to find new solutions.

What the TRIZ problem-solving process helps you do is bring together all these different approaches, get everyone communicating effectively and use and share the right knowledge to find the right solutions to the right problem.

Thinking functionally

Thinking functionally is a key skill that TRIZ helps you develop. Many people think functionally automatically, and in many technical fields this ability is taught explicitly. It's a useful method for uncovering the connection between what you want (benefits) with the real world (existing systems). Thinking in functions requires a more abstract way of looking at problems that's still very practical and useful.

Thinking functionally is at the heart of TRIZ, and Chapter 5 provides a good introduction to understanding functions. Chapter 6 shows you how to find new functions; the most powerful tool for problem solving is Function Analysis (covered in Chapter 12); and Chapters 13 and 14 provide you with the tools for developing and improving functions.

A *benefit* is an outcome that you want – with no description of how you get it. Thinking in functions is one step towards considering how you can achieve a benefit. Many ways of delivering functions exist and, as a result, you'll come up with many potential solutions.

Let's say the benefit you're looking for is a delicious cup of tea. Many functions are required to deliver this benefit; for example, you need to heat water, provide a container in which to hold the hot water, infuse the tea leaves in the hot water, remove the tea leaves from the tea, provide another container from which to drink the tea and supply the means to add milk, lemon or sugar, if desired. Lots of different solutions can provide these functions. Thinking about all the functions you want means that not only will you see more possibilities for new solutions but you'll also ensure that you capture all the requirements.

Starting Your TRIZ Journey

When you're first starting out with TRIZ, the easiest thing to do is pick up one tool and learn how to use it. Choose the one you think sounds most interesting. After that, add to your toolbox bit by bit. Each tool offers different benefits, and, while they do comprise a coherent step-by-step process (see Chapter 11 for details), you need to understand each tool individually before you can start putting them together.

Getting a handle on the TRIZ tools

Tools help you do a job, whether you're a carpenter, car mechanic, dressmaker or TRIZ wizard. Three different classes of TRIZ tools are available to help you achieve your goals:

- The tools based on patent analysis and scientific journals
- The tools developed to help you model your problem conceptually
- The thinking tools based on modelling thought processes

The tools based on patent analysis and scientific journals, capturing the clever solutions people have generated in the past in a conceptual form, are the:

- 40 Inventive Principles
- 8 Trends of Technical Evolution
- TRIZ Effects Database
- 76 Standard Solutions

The following tools help you model your problem conceptually, so that you can strip away unnecessary detail and access the right solutions to your problems from one of the four tools above:

- ✔ Contradictions
- ✔ Function Analysis
- ✔ X-Factor

Finally, come the thinking tools based on modelling the thought processes of the most creative problem solvers:

- ✔ Thinking in Time and Scale
- ✔ Ideal Outcome
- ✔ Resources
- ✔ Size–Time–Cost
- ✔ Smart Little People

Each of the tools and approaches has different benefits and will be more useful for certain types of problem.

Reapplying proven knowledge to deliver innovative new solutions

The problem-solving tools based on patent analysis and scientific journals – the 40 Inventive Principles, the Trends, the Effects Database and the Standard Solutions – are TRIZ's crown jewels. Even the idea behind them – to look at cataloguing known success – is incredibly clever and innovative.

Let's take a closer look at these crown jewels:

- ✔ The **40 Inventive Principles** (Chapter 3) are the clever ways of solving particularly hard problems: contradictions. When you have a problem that seems completely impossible, you probably have an undiscovered contradiction – you want two connected things which are in conflict. If you uncover and define the contradiction, the relevant 40 Inventive Principles will direct you to resolving the contradiction and finding new ways of getting what you want. The 40 Inventive Principles are also useful when you make a change to improve something but then, disaster, something else goes wrong as a result. You can also use them to develop and improve existing solutions.

✔ The **8 Trends of Technical Evolution** (Chapter 4) are useful when you want to develop, evolve and improve existing systems (products, processes or services). The Trends help you think about where you are now conceptually – and where you should be going. Some of the Trends are useful for understanding the maturity of your system and the best places to focus attention for future development; some are conceptual triggers of the likely future directions your systems will take. The Trends are particularly useful for developing next-generation systems, planning the future for your system development and for any patent or intellectual property work. The Trends are one of the easiest tools for people new to TRIZ to pick up and use immediately.

✔ The **Database of Scientific Effects** (Chapter 6) organises available online knowledge in a unique way to make it easily accessible, and is continually growing. Only so many ways have been uncovered for doing certain things, such as measuring weight or evaporating a liquid, and these have been captured and organised as simple 'how to?' questions and answers in this database of scientific and engineering effects. This list is useful when you want to know how to do something, usually something new; for example, when you're looking for a new function for a system or want to find another way of doing something because the ways with which you're familiar aren't good enough or are associated with big problems. The Effects Database is also useful when inventing, which is essentially looking for new functions. You probably won't use the Effects Database every day, but it's very useful when you do need it!

✔ The **76 Standard Solutions** (Chapters 13 and 14) are the ways in which you deal with three kinds of problem: a harmful action (something bad is happening that you don't want to happen); an insufficient action (something good isn't as good as you'd like it to be); or a need to measure or detect something (and applying conventional methods is hard or impossible). The Standard Solutions require a good understanding of how your system really operates on a functional level. They're useful for improving systems that don't have any big contradictions, so they are good for problem solving with mature and/or very complex systems with multiple interactions, particularly if you need to significantly reduce cost and complexity, where the Trimming Rules (Chapter 14) will be the most useful.

The fact that these tools have been distilled into relatively simple and easy lists so that successful innovations can be reused is very exciting, and one of the reasons why people often start describing TRIZ in relation to them (most commonly the 40 ways of solving contradictions). These tools were developed from technical problems and solutions; however, they needn't only be applied to these kinds of problems. Some of the same reasons that make technical problems hard to solve may also apply to other kinds of problems. For that reason, the clever solutions found can also be applied to those other kinds of problem.

These lists were derived from ideas created by people – not artificial intelligence or alien technology or divine revelation. One way of looking at them is as revealing patterns in human creativity rather than in technical innovation. Reapplying these technical innovations in other fields seems very – well – TRIZzy!

Modelling problems conceptually

In order to apply any of the tools mentioned in the preceding section, you first need to use TRIZ to understand your problems in a new way. The TRIZ tools for modelling your problems in an abstract way produce very powerful and clear thinking.

Follow the steps for implementing these tools and they'll guide you to understand and view your problem in a very different light.

These tools are:

- ✔ **Contradictions**, which are the problems that are getting in the way of you achieving everything you want. You put in checks to make sure a process is done correctly but then it takes longer. You want a big screen when you're reading something on your smartphone but a small device when it's in your pocket. What's important to bear in mind is that these contradictions only exist in the current ways of delivering what you want. People have always faced contradictions, and typically they reduce expectations and compromise, but every now and again very innovative people have found other ways of resolving contradictions that are so clever they're like tricks; they've broken out of the traditional way of thinking and found a really inventive new solution. All these solutions are summarised in the 40 Inventive Principles, and to access the solutions most relevant in your situation, you need to uncover and then define your contradictions. Doing so will allow you to break out of your traditional patterns of thinking and find new ways of getting what you want. Understanding contradictions is essential when you encounter really hard problems and just can't find a solution. They're also useful when you're inventing, improving imperfect solutions and encouraging creative thinking.

✔ **TRIZ Function Analysis**, which is the means of understanding your system thoroughly. You map all its current functions, identifying both what's good (useful actions) and bad (harmful, insufficient or excessive actions). You then have a list of problems, defined in a very clear way, and are able to use a number of the problem-solving tools, most commonly the Standard Solutions (Chapters 13 and 14) but also the 40 Inventive Principles (Chapter 3), if the Function Analysis has uncovered Contradictions. TRIZ Function Analysis is essential in any rigorous problem-solving work, because it uncovers and clearly highlights all potential problems and works best on systems that are real (rather than potential) and well understood, based on one snapshot in time, and it works on anything from new inventions to complex processes. It's useful for understanding the whole problem space and sharing that information within and across teams, for uncovering root causes of problems and charting complex situations. Function Analysis is essential in any system improvement work because it can be used to predict the impact of proposed changes and communicate both the situation as it is and how any new system would work.

✔ **X-Factor** (Chapter 6), which is one of the simplest innovation tools for modelling problems but forms the basis for accessing a number of the solution tools to help you find what you're looking for. When you define an X-Factor, you define the function which will solve your problem. Doing so means you both focus on what you hope to achieve and identify it in a way that's both very precise and completely independent of any current system or technology. This is important as it breaks your psychological inertia by starting with what you want and gives you a focused question to find the answer to in the Effects Database (Chapter 6), your resources (Chapter 5), the Standard Solutions (Chapter 13) or even a simple Internet search. The X-Factor is useful when you're inventing, improving a new system or looking for something you don't know how to deliver. It's particularly useful when you're dealing with smaller problems to which you need to find a solution quickly.

Supercharging your thinking

The simple TRIZ tools based on creative thinking techniques are powerful ways of shifting your thinking and developing your creative ability. They're a distillation of cleverness of a different kind to the solution tools.

When you think TRIZ, you start with what you want and then work out how to get it. As a fundamental philosophy that's very important, but it's also used as a formal tool in the form of the Ideal Outcome.

Here are the thinking tools available to you:

- ✔ **Ideal Outcome** (Chapter 9) is the means of capturing all the things you want. It makes you consider what you'd get if you could have everything you wanted. In terms of problem solving, you're looking to identify all the outcomes you want – all benefits, no solutions. This helps open your thinking to uncover all benefits, to think clearly, to challenge constraints on what's possible and to ensure you've set the right scope for your problem solving. The Ideal Outcome can be used as a very simple stand-alone tool for encouraging creative thinking, but it's also an essential early and practical step for every single TRIZ problem-solving or innovation session. It's of particular importance when you're attempting to create something new (for example, new product development) and to ensure team endorsement of goals and required outcomes.

- ✔ **Thinking in Time and Scale** (Chapter 8) is one of the quickest and simplest ways to think like a genius. Stretching your view of a situation to encompass not only your system but also the big picture and the detail, and how these three levels of scale are changing over time, will enable you to think with great clarity, see new connections, identify problems and ensure you're solving the right problem. When you've learned how to think in time and scale, it changes your thinking forever. It can also be easily used by people with very little (or no) TRIZ knowledge. Thinking in Time and Scale is another powerful tool for helping teams gain consensus on what's happening, understanding a problem and communicating it simply, and finding very innovative new solutions.

- ✔ **Resource thinking** (Chapter 5) is another tool that becomes a reflex as well as a formal tool. TRIZ thinking always pushes towards elegant self-systems to deliver what you want, and the easiest way to achieve this is via clever use of available resources. Using existing resources is particularly important in cost-saving situations and where strict regulations make it hard to bring in new technologies, components or substances, or changing the way things are currently done is extremely difficult. Resource thinking is also of particular importance when moving towards more sustainable solutions: everything about, within and around your system (even the problems) is made to work hard for you.

- ✔ **Size–Time–Cost** (Chapter 7) is an exaggeration thinking tool that challenges your perceptions of your constraints. People are often far more pragmatic about the solutions they suggest than they realise and what Size–Time–Cost does is stretch your thinking to the extremes but with some simple suggestions to direct you (can you guess how? The clue's in the name!). You imagine that your solution could be infinitely large or infinitely small – takes forever or works instantly – is subject to an unlimited budget or no budget at all – and then translate these notions into real terms. Size–Time–Cost is a very quick and simple tool for thinking creatively and generating unexpected (and often unexpectedly practical) new ideas.

✔ **Smart Little People** (Chapter 7) helps you both to understand your problem and find solutions, and to model both. You imagine your problem is made up of little people. Naughty ones come in and cause problems, and you capture what goes wrong as a result. Helpful little people then come in and solve the problems, and you translate this imaginary useful behaviour into concrete and practical solutions. Smart Little People is another powerful standalone creativity tool that can generate solutions very quickly, but it's also very powerful for breaking psychological inertia and allowing you to look at your situation from a completely different perspective.

Whereas the solution tools were derived from analysis of actual clever solutions (as described in patent records and scientific journals), the tools for creative thinking resulted from watching very clever and creative people at work. What was observed was that creative problem solving is the result of certain patterns of thinking. The TRIZ community detected these patterns and then codified them into formal thinking tools that everyone can put into practice, to think like a genius on demand.

One of the tricks many of the creativity tools employ is to stretch your thinking beyond the probable, or even the possible, into the realms of wild extremes. Do not resist this process! This step is designed to take you out of your comfort zone and into a new mode of thinking. The step *after* this wild thinking is to bring it back to reality. When you've become more familiar with the process, you can let your imagination fly with more confidence (because you've seen it work) and thinking in this way will come more naturally, and flexibly, until it becomes second nature to mimic this typical form of genius thinking.

Mastering TRIZ

So, you've got the basics under your belt. What do you do next to develop your TRIZ ninja skills? Read on!

Putting the tools together in the TRIZ problem-solving process

The most important thing you need to know about the TRIZ process for solving problems is that it's possible to have a process! And, more importantly, a generic process that works on any kind of problem.

The important problem-solving stages are:

1. **Understand and scope the problem.**
2. **Uncover all needs and scope the solution.**
3. **Zoom in and define the problem.**
4. **Identify the solution triggers.**
5. **Generate solutions to the problem.**
6. **Rank solutions and implement.**

As you can see, these steps are general enough to apply to any problem (which specific TRIZ tools to apply where is the subject of Chapter 11).

Applying the systematic TRIZ problem-solving process gives you:

- ✔ **Great clarity of thought:** The ability to uncover the heart of any problem you're solving and focus your attention on the right places.

- ✔ **Access to the right knowledge:** You find the solutions you'd have come up with anyway but then move beyond them by accessing the world's knowledge; not all of it – just the right, relevant, new knowledge to help you solve the problem at hand.

- ✔ **Innovative new solutions:** You make new connections between what you already know, and are given prompts for new ideas outside of your own experience. You thus find powerful solutions that you'd never have found without TRIZ.

- ✔ **Improved teamwork:** TRIZ helps people with diverse experiences and approaches work together well, drawing out the best in everyone's thinking and bringing it together in a coherent framework.

Most people are never taught problem solving as an explicit process except when tackling particular problems with well-known, specific, step-by-step processes such as cooking or solving differential equations. Generally, problem solving is one of those things that you're expected to pick up at work as you go along; you face a problem and work out how to deal with it, perhaps under the supervision of someone with more experience who can give you some guidance. This is how most people develop professional expertise: by encountering problems and finding solutions to them. Sometimes a problem occurs because something's gone wrong and you need to fix it or you haven't done everything perfectly the first time. According to Irish playwright and raconteur, Oscar Wilde, 'experience is the name everyone gives to their mistakes'; you learn as a result of solving the resulting problems.

As a result, it may seem that solutions to problems must be specific to the particular areas in which the problems occurred. For example, if a chemist has a problem, presumably only other chemists will be able to help, because they alone can understand the parameters of the problem and the nature of acceptable solutions. However, TRIZ tells you that, on a very fundamental level, there are simple rules for all problem solving and the nature of whatever problem you're facing will have been seen by someone else in the past. The TRIZ problem-solving process thus helps you step through your problem and strip out unnecessary detail to understand its nature more clearly and then to access the right solutions – which may well be outside of your industry. You access the right knowledge and can also involve other people in your problem solving.

Innovating with others by sharing solutions

Very few people look to develop new ideas or solve problems on their own. At the very least, other people are generally needed to help develop or implement innovations. As such, getting to know the TRIZ approaches that will facilitate effective innovative thinking with other people is important.

It's essential that everyone's able to share his ideas or solutions with others. You need to create a 'safe space' in which no one feels hesitant about voicing well-developed solutions, half-formed ideas, questions, comments and thoughts. The easiest way to do this is to call all solutions 'bad solutions', which simply means they're imperfect and capable of being improved upon (if you don't like the term 'bad', substitute it with 'initial' or something similar). This creates a mock humility about any solutions people come up with (everyone secretly thinks his solution is brilliant), and makes it easier for people to share solutions that they know to be imperfect or half-formed or even totally wacky.

Create a solution 'park' where everyone can see it – on a wall, for example. This way, everyone's solutions are visible, easily shared and of equal rank.

Cultivating the motivation for innovation

If you want innovation, generating new solutions isn't enough. Individuals, teams and organisations need to be motivated to innovate; they have to feel excited by the idea of improving things and actively look for opportunities to develop and improve the way things are done or even to do something in a completely different way.

Changing company culture and creating the right organisational processes for innovation are two massive topics, worthy of their own (*For Dummies!*) books. What TRIZ can deliver is changed attitudes to and beliefs regarding innovation, at an individual level (if someone's learned TRIZ); at a team level (if the manager supports it); and at an organisational level (if TRIZ has become part of the organisational way of doing things, as it has in several major companies such as Samsung). It's the individual level that interests me most because it's within the scope of control of you, the reader. Learning TRIZ encourages within you a different attitude towards innovation, problem solving and creativity. You're much more open-minded about what may be possible and have a 'can do' attitude towards problems because you know you can solve them. TRIZ also fosters persistence in the face of failure, as any roadblock to implementing something new is just a problem – that you can tackle with TRIZ! It also encourages a kind of restless energy that's the opposite of complacency in terms of seeking out problems and new places for improvement. Innovation is often not about the next big breakthrough product but a series of many small improvements to the way that you work and approach issues. You can implement these small improvements when you learn TRIZ.

Table 1-1 shows some of the helpful attitudes towards innovation that TRIZ fosters within individuals – and their opposites!

Table 1-1	Thinking About Innovation: The Wrong Way versus the TRIZ Way
The Wrong Way	*The TRIZ Way*
It's too hard to change things.	We can find new ways of working within existing constraints; we'll get everything we want without changing anything.
It'll never get approved.	It's always worth challenging constraints.
Things are as good as they're ever going to get.	Things can always be improved – we can increase their Ideality.
I don't know how to do something.	Let's define exactly what we need to do so we can find the right knowledge.
It's too big a mess to tackle.	TRIZ will help us understand the problem and define what we need to do.
We don't have time to do anything differently.	We'll find a quick solution.
We'll never find the answer.	TRIZ will help us find an innovative solution.

Being humble and looking like a genius

A lovely aspect of TRIZ is that not only does it help you make the best of your creative ability, but there's always something else you can try and the solutions you generate can always be improved upon. This both encourages you to keep working on and developing all solutions and makes ideas cheap (in a good way).

You can adopt one of two fundamental attitudes towards your ideas and solutions when you're working with other people: you can treat them as rare treasures, hold them close to your chest and only share them in exchange for large rewards; or you can give them away. Taking the second route not only implies that you're capable of generating huge numbers of new ideas but also actually makes it more likely to happen.

Now, I'm not suggesting that you give away your company's intellectual property portfolio in exchange for a handful of beans. Rather, I'm suggesting that when you're working with others, everyone will benefit from the free and frank sharing of ideas. Like love, the more you give away, the more you receive, because other people will share their ideas with you. Also, and this is important, when you let go of an idea, other ideas will occur to you (holding onto an idea blocks your thinking; nothing gets in the way of a great idea like a good idea).

This approach helps develop your sense of humility towards your own ideas, because you aren't expecting a fanfare and massive pat on the back every time you suggest something. Everyone's sharing, and everyone's ideas are regarded as valid, interesting and carrying some useful information (either about what you currently have, what you want or both). A sense of humility helps you work better with other people because you don't fight for your ideas to be recognised over those of others, and you're more open to your ideas being changed and developed by others. Incidentally, humility also makes you look like a genius, because sharing so openly clearly demonstrates that generating new ideas comes very easily to you!

Chapter 2

Understanding the Fundamental TRIZ Philosophy

. . . .

. . . .

*T*RIZ was developed to summarise useful solutions from analysis of previous engineering and scientific success, but the underlying philosophy regarding how to approach problem solving and innovation can be applied to many, many situations.

This chapter introduces some of the key concepts in TRIZ and its underlying philosophy, and helps you to 'think' TRIZ by respecting its origins and understanding how the method has evolved over time.

Above all else, as you read through this chapter, bear in mind that TRIZ thinking is achieved by considering two questions:

1. What do I want?

2. How can I get all the things I want without changing anything?

TRIZ tools can help you do this in a more formal way (see Chapters 5 and 7 for more about the relevant tools), but for experienced TRIZ people, this kind of thinking becomes another reflex – and encourages (and requires) a certain confident mindset.

Uncovering and challenging all your assumptions is important: following a logical process like TRIZ helps you do just that, and find the right solutions to your problems.

Thinking TRIZ

Understanding where TRIZ comes from helps to explain the logic behind it. This section explains how TRIZ started to be developed and how it can help you. TRIZ was developed to 'bottle' clever solutions that have been successfully applied to solve difficult problems and to do so in such a way that they can easily be reapplied in new situations. TRIZ was uncovered from real-world problems and solutions: when you learn how to 'think TRIZ' you are able to reapply other people's successful solutions to solve your problems.

Picturing TRIZ's beginnings

TRIZ was the brainchild of Genrich Altshuller (1926–96). Born in the former Soviet Union, he was a naturally very inventive person and as a teenager already had two Soviet equivalents to a patent to his name.

In 1946 Altshuller started work in the Soviet Navy's patent office and noticed a pattern in the patents he was studying: people tended to come up with very similar solutions to each other, but in different fields or different applications. With Rafael Shapiro, he began researching whether this was the case, and found that the same themes were repeated in particularly inventive solutions (check out the nearby sidebar, 'Solving contradictions: the definition of an inventive solution', for more info). Unfortunately for Altshuller, he fell foul of Stalin's political temper and was arrested, charged with 'inventor's sabotage' (a charge some clients tell me they think still exists in their companies!) and sentenced initially to death, then to 25 years hard labour in a Siberian gulag (to say it was a bad idea to disagree with Stalin is a bit of an understatement). In the gulag, Altshuller was surrounded by other political prisoners, including a lot of engineers and scientists. He discussed his theory with them and started to develop the early TRIZ community, right there and then. In 1953, after Stalin's death, many political prisoners were released, including Altshuller, Shapiro and other members of that community. They immediately started researching the theory that patterns exist in human creativity.

After analysing 50,000 patents, the early TRIZ community had uncovered 40 ways to resolve contradictions – the 40 *Inventive Principles*. These principles are very general; for example, they suggest doing things such as breaking an item into different pieces, changing its shape or making parts of it move. The global TRIZ community has continued researching but the number remains at 40 (although fans of *The Hitchhiker's Guide to the Galaxy* would like to up it to 42).

Solving contradictions: the definition of an inventive solution

What's an *inventive solution*? It's an idea that makes you say 'Aha, that's clever!' One of Altshuller's leaps of genius is that he realised a solution is particularly inventive when it solves a contradiction or resolves conflict. Human beings have always been solving contradic-tions, but what Altshuller and the early TRIZ community did that was so ingenious was to codify all the problems and answers in a generic way, so that when a contradiction arises, a shortcut is available to the useful answers other people have thought of in the past.

The first thing a lot of people new to TRIZ (particularly academics) want to do is find a new Inventive Principle. I've seen attempts to introduce new principles – one branch of the TRIZ community has suggested another ten – but these can always be seen as potential sub-sets of existing principles. In my experience, the 40 Principles work – they genuinely help people quickly find clever and inventive answers, and the time people spend uncovering and defining their contradictions results in much clearer problem understanding.

As soon as the 40 Inventive Principles had been uncovered, the TRIZ commu-nity started to analyse contradictions and categorise them. They then looked at which Inventive Principles resolved each particular kind of contradiction. As a result, when you face a contradiction, you can look up which of the prin-ciples have been used to solve one like it in the past. You can then use those principles to generate your own innovative ideas (see Chapter 3 for more on Contradictions and the 40 Inventive Principles).

After this first, important discovery was made, the TRIZ community contin-ued to analyse problems and solutions to create a knowledge base that, at Altshuller's insistence, was in the public domain for everyone to use.

Standing on the shoulders of giants – or reinventing the wheel?

Some people start to get a bit worried when they plunge into TRIZ. 'Hang on a minute,' they say, 'I'm really good at solving problems and being creative already. If I'm just reusing other people's inventions, doesn't that mean anyone can apply these skills and TRIZ is really boring and simply copying?'

My usual response to this is:

- ✔ Yes – everyone can do it! That's the point. Everyone can be creative – and even the most creative people can learn to be more creative. You're not relying on a random spark of inspiration or a stereotypical 'creative person' with mad hair and odd socks (not that there's anything wrong with these features, of course!).

- ✔ Yes – it is systematic but it's also fast and flexible! So rather than reinventing an existing solution or product, you can put all your energy and cleverness into developing an idea further and making it work in new applications. Most clever inventions involve applying existing technology in a new field or new combinations. Jeff Bezos, for example, founder of Amazon, didn't invent the Internet or bookshops. Rather, he put them together to create an innovative new company.

- ✔ No – you're not copying! Because the nature of TRIZ solutions is very general, and you have to apply them in a specific context, you're not acting like a human Xerox machine, you're being inspired. The inspiration derives from a useful and proven solution – but it still requires a lot of creativity, clever thinking and domain knowledge to turn it into something practical and useful. This is the most exciting part of using TRIZ – your brain is thinking freely, you're making new connections between existing knowledge and you're creating something completely new.

Understanding the TRIZ Philosophy

Underpinning the TRIZ tools and processes are approaches to thinking that help you see things more clearly, recognise new possibilities and become a better problem solver. These thinking skills are not only at the heart of the TRIZ logic but are also useful habits to develop if you want to reliably think like a genius – even when you're not explicitly applying TRIZ tools.

Learning to think conceptually

Thinking conceptually is one of the most important skills you can learn from TRIZ. It means discovering how to strip out the detail of a situation and see it in a more general, abstract way. Thinking conceptually enables you to see many new possibilities and stops you getting stuck in unnecessary detail and developing psychological inertia in relation to a situation or solution (see Chapter 7 for the lowdown on confronting psychological inertia).

To be able to think conceptually, you need to understand the difference between a concept and an idea. A *concept* is a general way of doing something; an *idea* is a specific way of putting that concept into practice. For example, 'cook an egg' is a concept; 'frying an egg' is an idea. Boiling, poaching and scrambling are also ideas – you can cook an egg in many, many different ways!

When you've identified the general concept of thinking conceptually, it becomes very easy to see other ways of putting that concept into practice, and to generate many other ideas.

Having learned how to look at the concepts behind each idea proposed by other people, it's also very easy to build on and improve their ideas or suggest alternative ways of putting their concept into practice. When someone says, 'Oh but you can't do that because . . .' you can suggest an alternative idea resulting from the same concept. This approach helps you think more flexibly and creatively. Generating ideas is much easier when you have a starting point – and the concepts behind other people's ideas can provide just that. Even if you can't think of an idea yourself, you can use intelligent questions to prompt others to come up with different ways in which to put that concept into practice.

Idea–Concept is a simple creativity tool to help people generate lots of ideas. It is very useful when you need to generate a lot of ideas quickly, and works both when you are working alone and with other people. What usually happens when you are given a problem is that you often find it easy to think of one or two ideas immediately; however, you can quickly run dry. Always allow your initial free thinking and natural spark of creativity to happen first, and capture those initial ideas. However, when you are running out of steam, you need to apply an explicit tool such as Idea–Concept. When lots of ideas are needed, teams are often brought together to brainstorm: using Idea–Concept will help supercharge any team brainstorming activity.

When using Idea–Concept with a team:

1. **State the problem.**
2. **Ask everyone to think of ideas individually and write them down.**
3. **Gather the ideas together and identify the concepts that lie behind them.**
4. **Take each concept and use it to trigger new ideas.**

Consider, for example, asking people to come up with as many uses of an object (such as bubble wrap) as possible. This is a simple way to get people to understand the difference between ideas and concepts. Figure 2-1 shows an example of how this works.

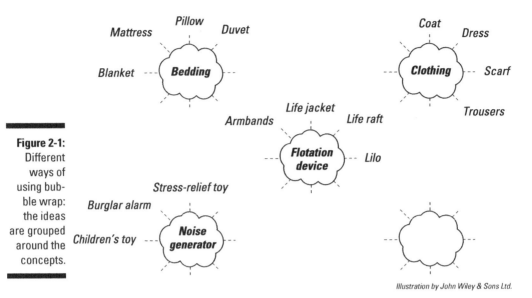

Illustration by John Wiley & Sons Ltd.

Figure 2-1: Different ways of using bubble wrap: the ideas are grouped around the concepts.

Using Idea–Concept thinking, going from 3 initial ideas to 30 very quickly can be pretty easy. Bear in mind that some clever ideas have more than one concept behind them. That's okay; it just gives you more concepts to play with!

Creativity is difficult to measure. The Alternative Uses Task, devised by psychologist J.P. Guilford (1897–1987) in 1967, is one way in which to do so (and has been used in one form or another for most creativity tests since). It asks you to list as many different ways to use a brick as possible as a means of measuring your divergent thinking. The number of ideas you come up with is one measure of your creativity, and flexibility, originality and elaboration are also gauged. I had to complete this task at school and in more than one university interview; if I'd known about TRIZ and Idea–Concept thinking back then, I may have come across as a genius!

Reapplying proven knowledge to solve problems

The TRIZ approach says that any problem you encounter has probably also been dealt with by other people. If you can learn to look at both problems and solutions in a more general way, you can see how your problem is similar to other people's problems, and then work out how to reapply their clever solutions to your situation.

When you start to think in a more abstract way, it becomes easier to see how to reapply proven knowledge. In TRIZ, this approach is described as moving around the *Prism of TRIZ*. Check out Figure 2-2 for a funky diagram of this concept.

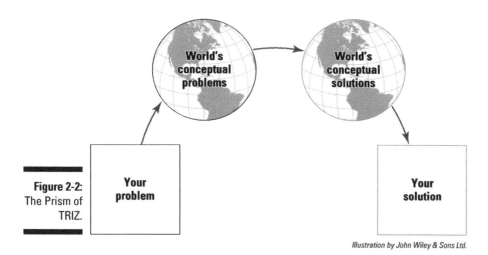

Figure 2-2: The Prism of TRIZ.

Illustration by John Wiley & Sons Ltd.

James Dyson used cyclone separation to create his bagless vacuum cleaner, using technology that separates dust from air utilised in the sawmill industry for more than 100 years. De Mestral invented Velcro after noticing how burrs attached to his dog's fur. Electric car manufacturer Tesla copied Apple's business model of placing company-owned retail outlets in prestigious locations rather than the traditional approach to selling cars via franchised dealers. This was a clever analogy to draw on, because Tesla (like Apple) sells on brand rather than product.

TRIZ tells you that if you want to be a genius problem solver and to develop innovative ideas, you don't need to solve every problem from scratch – you can use existing knowledge to generate many clever and inventive solutions. As soon as you learn to strip out the detail, you can then go and look, systematically, for useful concepts that will solve your conceptual problem. Adopting this approach – asking whether anyone has solved this problem before – will help you use your creativity most efficiently and develop your inventive problem-solving ability.

When looking for similar problems in different industries, start with the people for whom this problem is life and death, those working in industries that operate in difficult conditions such as below sea or in space (NASA actively promotes the reuse of its technology – check out `http://technology.nasa.gov/` for details of NASA's Technology Transfer Program, which has a number of great online resources such as a searchable patent portfolio, success stories and an inventions and contributions board). Other good places to look are in industries that spend lots of money researching and developing solutions, such as big pharmaceutical companies, or those that produce goods in very large quantities because they'll have worked very hard to make their systems reliable and cheap. Patents are also a great place to start: the point of a patent is to describe how a new invention or innovation works, but many have expired or lapsed, meaning they are available for use – check out Google Patents (`https://patents.google.com/`) and Espacenet (`http://worldwide.espacenet.com/`).

Getting everything you want without changing anything

The statement 'getting everything you want without changing anything' contains two fundamental elements of the TRIZ philosophy.

The first is, go for what you want even if problems exist. TRIZ problem solving is positive: you start by identifying all the things you want, in the belief that you'll be able to achieve them. Sure, you'll encounter problems, but you'll be able to solve them (see later in this section for more on this idea).

The first thing that happens when you start listing all the things you want is other people telling you all the reasons why it's impossible, impractical, too expensive, too time-consuming and so on . . . however . . . the TRIZ philosophy says that you somehow get everything you want without changing anything – so you don't incur any extra cost, have to make the effort to change your system or face any risk. This TRIZ statement sounds really wacky. And it is. And that's why it's so useful – it forces you to stretch your thinking, uncover the things you want and then find inventive and resourceful solutions to the problem. I've solved countless problems using this philosophy, both in my own working life and at home, on everything from hiring new staff to devising new workshops. Thinking in this way is useful because it's the opposite approach to the pragmatic thinking most people employ day to day. If you're producing a new version of a product, it's very easy just to look at the existing product and tweak it slightly – starting from what you have. In contrast, this approach suggests that you start by thinking about what you'd offer in an ideal world – what you really want – and then identify how it can be achieved without changing anything. Sometimes this ideal scenario is possible; at other times change is required but often on a much smaller scale than you imagined. (Feeling thirsty? Check out the 'Share a Coke' sidebar nearby for a great example of a small but effective change.)

Share a Coke

Coca-Cola's recent 'Share a Coke' marketing campaign was highly successful. It simply replaced the Coca-Cola logo with popular names, such as Chloe and Harry, to encourage people to connect with the brand. What's so clever about this campaign is that the company found a way to personalise its product for its customers by changing very little: the actual product remained the same, all that changed was one word on the packaging. While this did require a lot of work – a new font had to be created (to address trademark issues), names chosen and approval granted – it was considerably less work than creating a specific product for each customer. The campaign clearly couldn't reach absolutely everyone – only the most popular names were featured – but Coca-Cola addressed this issue by encouraging consumers to 'share', which meant they could buy a drink for a friend if they didn't see their own name.

Developing confidence

Confidence helps you take your problem solving in the right direction. TRIZ encourages you to go for what you want even if there are downsides, because you can be confident you'll get there in the end by solving problems as they occur. This confidence means you're more likely to produce real innovation because you leap forward to what you want rather than shuffle along to some pragmatic quick fix.

Confidence breeds perseverance, which is an important element of creative problem solving. When you put something new into practice, inevitably, alongside the benefits, you begin to identify other things that aren't going to plan or encounter unexpected problems. What do you do then – give up? No, you say 'Great, another problem', and tackle it with TRIZ. Being confident that you can deal with anything thrown your way means that you don't give up when things get difficult or go wrong. Instead, you learn from, fix and improve them.

Confidence is crucial to the thought processes that encourage creative thinking. Research demonstrates that people who learn TRIZ are more confident in their ability to generate creative ideas, which often comes as a surprise to engineers, scientists and so on. With TRIZ in your mental toolbox, you too can feel confident in your ability to tackle any problem using creative thinking.

Confidence is a necessary but not sufficient condition for creative problem solving: you have to believe that you can solve a problem before you even begin to tackle it – but you still need the right tools to do so.

Confidence comes from within. But when you're anxious you can still put your trust in TRIZ. I've facilitated numerous problem-solving sessions in which people tackle large, difficult, seemingly insurmountable problems with serious downsides if they remain unresolved (plummeting share prices and redundancies, for example). In those scary moments I've seen teams of people clutching their matrices and charting their progress through the problem-solving map. They didn't need to have confidence in their own cleverness: the TRIZ process would take them in the right directions to good solutions. And this knowledge let them let go, stop panicking and start thinking clearly.

Being Systematic and Creative

Being creative requires more than simply unlocking your brain. To really solve problems cleverly – and work well in teams – you need a systematic process. Following a logical process ensures that innovative solutions are generated reliably and on-demand, you don't end up down any creative cul-de-sacs, and in fact your thinking is much more free because you don't have to plan what you're going to do next or why. You use the professional expertise and intuitive skills you have developed but challenge their boundaries: using them as a springboard for new thinking.

Using TRIZ processes for understanding and solving problems

Stories abound of great scientific or engineering discoveries being made in the shower, on a bus or in a dream. Revelations strike like a bolt from the blue. What a lovely experience. But how completely useless to you when you're at work, something's gone wrong and you need to fix it *now*!

When a piece of equipment starts leaking, engineers can't announce they need a nice long walk or a little nap to think about the problem and hope for inspiration. What usually happens is that people sit down together and try to understand logically what's going on and then find a solution. Often this scenario involves a meeting or brainstorming session, but what is even more useful at this point is using a systematic and reliable process to ensure the real problem has been uncovered and the right solution is found.

Using logical processes for understanding and solving problems ensures that you cover all the necessary parts of the issue, identify the potential problems and then generate solutions – reliably. This approach is less dramatic than random inspiration (and potentially involves less fun) but it's a lot more useful!

TRIZ step-by-step processes for problem solving also help keep everyone engaged together, confident of their direction and able to work through all the necessary information. Some people like to jump straight to brainstorming – other people like to work through a full and detailed understanding of the problem. Both are useful and reveal important information, so it's important that both are completed: having efficient systematic processes which take you to the right, relevant solution spaces ensures that there is time enough for both.

In stressful situations, it's tempting to just go with the first good solution that's suggested. However, that solution may be premature and you may miss some crucial part or cause of the problem. Logically stepping through the problem ensures that the whole problem is covered and the right solutions uncovered.

Thinking freely

Perversely, thinking in a structured way helps your brain think more freely. Many problems are very complicated and involve taking in a huge amount of information. TRIZ helps you chunk down the information into manageable pieces, and shows you where to focus your attention. Really difficult and complex problems can also be a bit intimidating; it can be hard to know where to start, and taking in the whole thing all at once can feel overwhelming.

Working through the TRIZ process means that when you're trying to think of new solutions, you do so one step at a time. Rather than trying to solve the whole problem all at once, you deal with small sub-problems one by one. This structure helps you think completely freely at each step – and you can bounce around in your thinking as much as you like – so long as you come back to the process eventually.

Challenging assumptions

The thing about assumptions is that you usually don't know you're making them. Your assumptions are sometimes implicit and sometimes the result of hard-won experience of what works and what doesn't. Someone once told me that he didn't need TRIZ as he'd seen every problem that could possibly occur in his industry and knew the solution to it. This kind of experience and expertise is fantastically useful for efficient and quick problem solving. However, it also means that the same problems are always fixed with the same solutions, and perhaps other (better, cheaper, faster) resolutions exist.

Psychological research into expertise shows that experts are able to *chunk* information together and, over time, identify patterns in what works and what doesn't. Ultimately, they're able to remember the patterns rather than the individual pieces of information. For example, this allows expert chess players to make clever moves because they recognise certain patterns and what to do next. Such pattern recognition can become automatic in any area of expertise. I've been singing in choirs all my life and can sight-read well because I don't need to work out the notes one at a time – a line of music comprises three or four patterns put together that I can recognise at a glance.

Professional intuition evolves just like that. Experts are able to identify problems and solutions quickly and with great ease because their experience highlights the patterns and key features of the situation, and they almost instantly can see a solution. Getting to this point requires a vast amount of practical experience, but allows for incredibly fast thinking once it's been acquired. This ability is crucial in highly stressful jobs that often involve time-critical problem solving, such as nursing in intensive care, flying a plane, working in air traffic control or managing a crew of firefighters. Fortunately, most people have a little more time to reach a solution. And thinking time can be quicker in the long run than implementing the wrong solution. Intuitive thinking suggests the right solutions nearly every time but only in routine situations.

What if the situation isn't routine? And what if you need an innovative solution? This is the point at which you need to challenge your thinking and look for new solutions, but the more experience you have, the harder this will be. So, allow your expertise to suggest the nature of the problem and potential solutions, but then apply an explicit method to challenge your thinking, investigate the problem further and look beyond your intuition to find new solutions.

Another kind of assumption concerns what's possible. The world is always changing – and perhaps the conditions you assume exist or the limitations of a certain technology just aren't relevant anymore. Sometimes very promising solutions are rejected too soon because people think they're impossible. However, just because you don't know how to do something, doesn't mean that no one else does. It's always worth finding out if the knowledge you seek is available in your own company.

Often teams don't hold the same assumptions either, about the type of solution they're looking for, the nature of the problem or the best way to go about tackling it. Consider my favourite joke: an engineer, a physicist and a statistician are at work one day when suddenly a wastepaper basket catches fire. The engineer starts looking for an extinguisher in order to put out the fire, the physicist begins calculating exactly how much thermal energy is being produced in order to plan the next step and the statistician sets fire to all the other wastepaper baskets in the room. 'What are you doing?' say the engineer and physicist in alarm, to which the statistician replies, 'We need a bigger sample.'

Part II
Opening Your TRIZ Toolbox

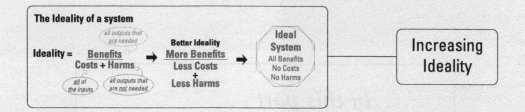

The Ideality of a system

Ideality = $\dfrac{\text{Benefits}}{\text{Costs} + \text{Harms}}$

all outputs that are needed

all of the inputs

all outputs that are not needed

→ **Better Ideality**
$\dfrac{\text{More Benefits}}{\text{Less Costs}}$
+
Less Harms

→ **Ideal System**
All Benefits
No Costs
No Harms

Increasing Ideality

web extras

Head to www.dummies.com/extras/triz for a free article that summarizes how to invent something using TRIZ tools.

Part II

Opening Your TRIZ
Toolbox

In this part . . .

- ✔ Find and solve contradictions.
- ✔ Develop next-generation systems with the Trends of Technical Evolution.
- ✔ Use resources inventively to improve your system's Ideality.
- ✔ Reapply other people's clever solutions with the TRIZ Effects Database.

Chapter 3

Solving Contradictions with the 40 Inventive Principles

*T*his chapter provides an introduction to one of the most popular TRIZ tools – the 40 Inventive Principles – and the powerful approach to understanding and solving your most difficult problems – contradictions. Contradictions are the kind of problems that most people avoid, because their resolution sounds impossible: when you have a contradiction you have a conflict; for example, 'I want a refrigerator that holds lots of food but takes up very little space in my kitchen'. The pragmatic and apparently sensible approach is usually to pick one thing over another (big or small) or find some kind of compromise (a medium-sized refrigerator). TRIZ problem solvers say 'No – I'm going to get everything I want and resolve the contradiction (all my food at the right temperature and a nifty kitchen)'.

Uncovering and Understanding Unresolved Conflicts

Have you ever seen a problem so hard that finding a solution seems completely impossible? If so, you've probably been wrestling with a contradiction. This section gives an overview of what contradictions are, how to spot them and what to do next.

Spotting contradictions and seeing how they're resolved

A contradiction is when you have conflicts in what you want: either you want opposites of the same thing, or as you improve something, something else gets worse. Contradictions exist everywhere: they've always been around, and human beings, being natural problem solvers, have found clever ways to resolve them.

Some everyday examples of contradictions include:

- ✔ I want a cup that keeps my coffee hot but doesn't burn my hand.
- ✔ I want a more powerful engine for my car but I don't want it to get heavier.
- ✔ I want to cover a large surface area quickly when I'm painting but I don't want to make a mess.

Take an umbrella. What you really want from an umbrella are opposites: you want it to be really big when you want protection from the elements and really small at all other times. Making an umbrella collapsible is the commonest way to resolve this contradiction: it can then fold down to become small enough to carry around but also open out to become big enough to shield the user from the rain. This is a very old solution; collapsible umbrellas have been around for thousands of years (you can find descriptions of them being used in Ancient Greece and dynastic China). One theory is that this design was inspired by tents, which have the same conflicting needs: providing shelter and being portable. Figure 3-1 shows how the conflicting needs of both objects are resolved.

The reason a tent is a clever invention is that it resolves the conflict of being both big and small. If you only want big, the problem isn't hard to address – build a house! What's difficult is creating a dwelling that you can also pack up in a bag and carry around with you.

If it's true that umbrellas were inspired by tents, this is a good example of problem solving by analogy. Who's dealt with a similar problem in the past and how can you reapply his solution? This is efficient problem solving: you put your knowledge and experience into making a known concept work for you.

TRIZ takes problem solving by analogy a step further. Having something both big and small is one kind of contradiction – many others exist. The TRIZ community recorded and catalogued all the different kinds of contradiction and people's solutions to them in the past. You then have to apply these very general solutions to make a practical invention.

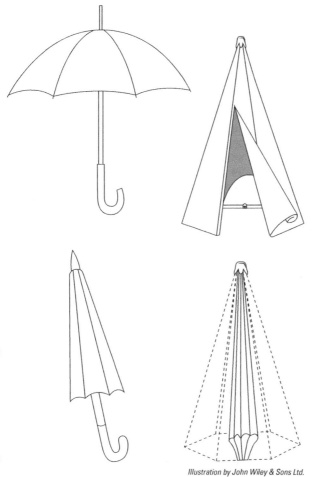

Figure 3-1:
Both
umbrellas
and tents
are
collapsible.

Illustration by John Wiley & Sons Ltd.

TRIZ identified just 40 ways of resolving contradictions: the 40 Inventive Principles. These are simple and easy-to-understand conceptual solutions, which suggest ways of changing your system to get what you want – in clever and inventive ways. I explain these principles in the later section, 'Reapplying other people's genius solutions'.

Understanding the two types of contradiction

Two types of contradiction exist – technical and physical:

- ✔ **A Technical Contradiction** is when, as you improve something, something else gets worse. You start with a solution but when you put it into practice, you find a downside. For example, you make a car safer by using lots more material but it makes the car heavier.

- ✔ **A Physical Contradiction** is when you want opposites of the same thing, like an umbrella needing to be both big and small.

The way that you get everything that you want (a safe and sensibly sized car or a big yet small umbrella, for example) is to understand you can separate the things that you want from each other. The traditional approach to contradictions is to compromise or give up; the TRIZ approach says other ways exist, but first you have to understand clearly what you want, and know it's possible to get it.

Contradictions are particularly hard problems and are often behind issues you struggle to find solutions to. Contradictions often seem impossible or, at the very least, extremely difficult when you describe them – and hard problems, by their nature, require clever and inventive solutions. So, how do you resolve contradictions and get everything you want? By using one of the 40 Inventive Principles.

Reapplying other people's genius solutions

The 40 Inventive Principles are all the clever ways people have come up with to resolve contradictions in the past, as recorded in patents. The TRIZ community back in the 1950s and 1960s identified that only about 20 per cent of patents were really inventive – the rest were relatively small or obvious changes – and what made these patents inventive was that they managed to solve a contradiction. The question then became: how many ways are there to solve a contradiction? Beginning with Genrich Altshuller and Rafael Shapiro, the first researchers to start developing TRIZ, and continuing with a community of engineers and scientists, a huge amount of research was conducted. The researchers analysed inventive patents and the contradictions they solved, and also, on a conceptual level, the solutions inventors had used to resolve contradictions. After they'd analysed 35,000 patents, they'd found 37 solutions; when they got to 50,000 patents, they'd found 40. These are the 40 Inventive Principles.

Are there really only 40?

The very first question almost everyone asks when they hear about the Inventive Principles is, 'Are there really only 40?' I remember my own incredulity – that can't be right; there must be more! Well, research continues (more than 3 million patents analysed so far) and the number still holds and that's because the 40 Inventive Principles are, by their nature, very general. They form one of the 'golden rules' for problem solving, like all the TRIZ solution triggers (including the Trends of Technical Evolution, the Standard Solutions and the Effects Database).

These 40 principles are very general. In reality, while the 50,000 patents contained 50,000 clever ideas, many similarities existed between them – at a conceptual level. The genius of the analysis is that it uncovered these basic principles – the concepts – and distilled them into a simple list (you can find the full list, with examples, in Appendix A).

To use the 40 Inventive Principles you have to be aware of the distinction between a concept and an idea. In TRIZ-speak, a *concept* is a general way of doing something and an *idea* is a specific way of putting that concept into practice. For example, take 'nesting' as a concept (Inventive Principle 7; see Appendix A): different ideas suggested by nesting include nested tables, tea nested inside a tea bag, nested measuring cups and so on.

Because the Inventive Principles are conceptual, you can reapply them without simply copying. They really are 'principles': general ways to do things or rules. The Inventive Principles suggest solutions such as' make something porous', 'make it curved' or 'do something in advance'. The really clever bit is to work out how these principles could be useful in your specific situation. You aren't reinventing the wheel: you're finding out how you can use a wheel to help solve a problem.

What the TRIZ community uncovered was a way of modelling contradictions and their solutions in a more conceptual way, and to solve contradictions you have to step through what's called the *Prism of TRIZ* (explained fully in Chapter 6, but shown here in Figure 3-2). To do this, you distil your real-world problem to a more simple, conceptual problem: this is your contradiction. You can then look up the Inventive Principles that will be most useful for you, and use those concepts to diverge your thinking and generate practical real-world solutions. When you step through the Prism of TRIZ' you leave the real world and think about your problem in a more conceptual, abstract way; that is, TRIZ suggests a conceptual solution from which you can generate lots of specific ideas.

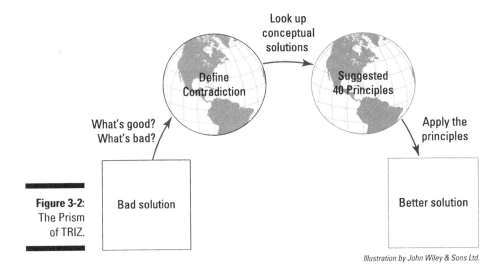

What's good?
What's bad?

Look up
conceptual
solutions

Define
Contradiction

Suggested
40 Principles

Apply the
principles

Figure 3-2:
The Prism
of TRIZ.

Bad solution

Better solution

Illustration by John Wiley & Sons Ltd.

Solving Technical Contradictions

When something improves and something else gets worse, you have a Technical Contradiction. Technical Contradictions can be there as an inherent feature of your system, but they're easiest to spot when you change something. A classic engineering contradiction is between strength and weight: as you make things stronger, they also tend to get heavier; by the same token, if you make them lighter, they tend to get weaker. This is typically the case because to make things stronger you use more material: a tank is stronger than a car partly because it's made of a very large amount of metal. However, what TRIZ tells you is that a car only gets heavier because you're using that particular, commonly used solution – more material. You typically think of strength and weight as being connected, as shown in Figure 3-3.

The first step to solving a Technical Contradiction is spotting that you have one – because it isn't always obvious! You first need to ask yourself whether you can get all the things you want. If you can, your problem is solved and you can put the solution into practice. If, however, getting something you want results in further problems, you probably have a contradiction.

Creating something very strong but also very heavy isn't a clever outcome. What you need to do is find another way – possibly beyond your own knowledge – of getting both those things. What the TRIZ community identified is that when someone's created something extremely strong while also very light, they've used an inventive solution. The inventor managed to solve the contradiction in a clever way, often by bringing in knowledge from another industry, technical field or area of science. Solutions such as these were captured as Inventive Principles. The TRIZ community then catalogued all the Technical Contradictions they saw being described and solved in patents in the Contradiction Matrix (which I explain shortly). You can now look up how people solved Technical Contradictions in the past and use their clever answers to solve your problems.

Figure 3-3: You encounter contradictions because you assume the things you want have to be linked – TRIZ shows you how to break that connection.

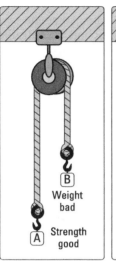 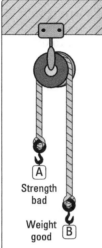 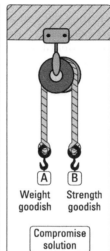 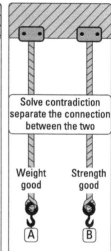

Illustration by John Wiley & Sons Ltd.

The five levels of invention

The Inventive Principles aren't based on all patent records. What the TRIZ community first identified was that not all patents are clever. In fact, patents were ranked according to five levels of invention:

> Level 1: solutions found within a narrow speciality (32 per cent of patents)

> Level 2: solutions found within the industry (45 per cent of patents)

> Level 3: solutions found in other industries (18 per cent of patents)

> Level 4: solutions found in science, mostly physical and chemical effects and phenomena that aren't commonly used (4 per cent of patents)

Level 5: solutions found beyond the boundaries of contemporary science; first scientific discoveries were made, then this new knowledge was used to solve the problem (< 1 per cent of patents)

The TRIZ community noted that finding Level 1 and Level 2 solutions isn't hard – most people should be able to do it, often by trial and error. Finding Level 3 and 4 solutions – deriving from other industries or applying scientific concepts that aren't commonly used – is inventive. Finding these solutions through trial and error is hard because they require a leap of imagination. The Inventive Principles are drawn only from the clever and inventive solutions found in Levels 3–5.

Improving imperfect solutions with a little help from your friends

Spotting Technical Contradictions when you're working with other people is easy, because the first thing that happens after you suggest an idea to someone is they tell you what's wrong with it! When someone says 'that won't work because. . .' they're pointing out a Technical Contradiction. This problem-identifying approach to other people's ideas seems to be a fundamental part of human nature: we all love our own ideas and can't see anything wrong with them; other people's ideas, however, seem full of problems! This kind of logical, problem-finding thinking is fantastically useful but not at the initial conception of an idea: at that point, it's best to allow ideas to flow without judgement. Then, if you can capture all the problems associated with that idea, you can see them as opportunities to develop and improve on it: you can solve those problems and improve your idea very early on in the process.

Using the Contradiction Matrix

When you've identified a Technical Contradiction, you can resolve it by using the *Contradiction Matrix*. This matrix resulted from the patent research undertaken by the TRIZ community in the 1950s and 1960s. Technical Contradictions can be mapped onto 39 Engineering Parameters, shown in Table 3-1, such as strength and shape. When these parameters had been identified, researchers classified inventive patents that resolved Technical Contradictions according to these parameters; then recorded which Inventive Principles had been used to solve the contradiction, and how often each Principle was used. Performing statistical analysis then allowed them to build up a matrix of the most commonly used Principles that solved each specific contradiction.

You can look up your contradictions in the Contradiction Matrix in Appendix B and see which Inventive Principles, described in Appendix A, are most likely to be useful for you.

For example, you want to make a cheese grater that can grate cheese really quickly. A simple solution is to make it bigger. However, if you make it really big, the cheese goes all over the kitchen. These two parameters are fundamentally linked: as the size of the grater decreases, control over where the cheese goes increases, but the rate at which you can grate cheese drops. A conventional approach to this problem is to find a reasonable middle ground, a compromise: a size of grater that allows fast-enough grating and a reasonable-enough ability to direct cheese.

Table 3-1   The 39 Parameters of the Contradiction Matrix

No.	Title	No.	Title	No.	Title
1	Weight of moving object	14	Strength	27	Reliability
2	Weight of stationary object	15	Duration of action by a moving object	28	Measurement accuracy
3	Length of moving object	16	Duration of action by a stationary object	29	Manufacturing precision
4	Length of stationary object	17	Temperature	30	External harm affects the object
5	Area of moving object	18	Illumination intensity	31	Object-generated harmful factors
6	Area of stationary object	19	Use of energy by moving object	32	Ease of manufacture
7	Volume of moving object	20	Use of energy by stationary object	33	Ease of operation
8	Volume of stationary object	21	Power	34	Ease of repair
9	Speed	22	Loss of energy	35	Adaptability or versatility
10	Force	23	Loss of substance	36	Device complexity
11	Stress or pressure	24	Loss of information	37	Difficulty of detecting and measuring
12	Shape	25	Loss of time	38	Extent of automation
13	Stability of the object's composition	26	Quantity of substance/the matter	39	Productivity

The TRIZ approach is to say, 'No – I want both: I want incredibly fast cheese grating but also to be able to direct the cheese to where I want. How can TRIZ help me get both?' You can use the Contradiction Matrix to tell you how people who've faced similar problems in the past solved this contradiction. Now, before you get too excited, the Contradiction Matrix doesn't mention cheese! In order to use the Matrix you have to think about your problem in a more general way – stripping out any detail about your specific situation. You first understand what your contradiction is, and then you see which of the 39 Technical Parameters your contradiction fits against.

Contradiction: *Faster cheese grating versus bigger surface area.*

Now you have to see how what you've identified as good and bad matches with the 39 Parameters. I've done it for you, and the results are shown in Table 3-2.

Table 3-2	Uncovering Your Technical Contradiction
What's getting better?	**What's getting worse?**
Faster cheese grating	Lost cheese
TRIZ Technical Parameter: 39. Productivity	TRIZ Technical Parameter: 6. Area of stationary object

Now turn to Appendix C, where you can see the 39 Parameters laid out in the Contradiction Matrix. If you look up where 39, Productivity, intersects with 6, Area of stationary object, you find the numbers: 10, 35, 17, 7, as shown in Figure 3-4.

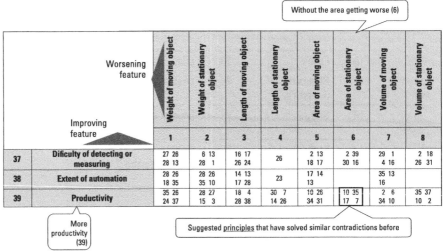

Figure 3-4: Looking up this contradiction in the Contradiction Matrix.

Illustration by John Wiley & Sons Ltd.

These numbers are the 40 Inventive Principles that people have used most often in the past when trying to solve a problem, in this case: as productivity increases, the area gets worse. (I cover the Principles in more detail in the next section.)

The next step is to take each of these Principles and work out how to use it to generate a practical solution. So, taking each in turn:

10. Prior Action. This would suggest buying pre-grated cheese or spreading newspaper to catch the errant cheese.

35. Parameter Change. One parameter that can be changed is the degree of flexibility, so create a flexible cheese grater that can be big to grate cheese but curve to direct cheese where you want it.

17. Another Dimension. Go from two to three dimensions by curving the cheese grater, and also increasing the length, which gives a larger surface area for grating but directs the cheese, such as the grater shown in Figure 3-5.

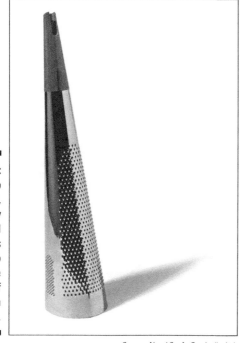

Source: Alessi S.p.A., Crusinallo, Italy

Figure 3-5:
The Todo giant grater, designed by Ricard Sapper, was created to grate one portion of cheese with one stroke.

7. Nested Doll. Place an object inside another – so you have a built-in container for the cheese underneath the grater, as shown in Figure 3-6.

What you're doing is stepping through your *Prism of TRIZ* (graphically demonstrated in Figure 3-7) from your real-world situation to understanding your problem in a more general way (Chapter 6 covers the Prism of TRIZ in more detail). You can then see how it's similar to other people's problems, and take the clever solutions from the past to apply to your situation. Familiarise yourself with the 39 Parameters and 40 Principles and apply your creativity and experience and you have a very sure route to generating practical solutions.

Figure 3-6:
The
Parmenide
cheese
grater,
designed by
Alejandro
Ruiz for
Alessi,
stores the
cheese
inside the
grater.

Source: Alessi S.p.A., Crusinallo, Italy

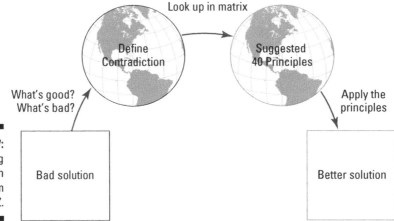

Look up in matrix

Define Contradiction

Suggested 40 Principles

What's good? What's bad?

Apply the principles

Figure 3-7:
Stepping
through
the Prism
of TRIZ.

Bad solution

Better solution

Illustration by John Wiley & Sons Ltd.

So, to uncover and solve contradictions, follow these steps:

1. **Identify your Technical Contradiction (a solution makes something better but causes something else to get worse).**

2. **Ascertain which technical parameters these relate to (use the table of technical parameters in Appendix C).**

3. **Look up where the technical parameters intersect in the Contradiction Matrix in Appendix B.**

4. **Use the Inventive Principles suggested to explore solutions.**

Getting to Grips with Physical Contradictions

A Physical Contradiction occurs when you want opposites of the same thing. I want a hypodermic needle to be sharp and blunt: sharp when I'm injecting myself with insulin, and blunt at all other times (so it can't hurt me or other people). I want a sticking plaster to be sticky to attach to my skin, but not sticky where it's covering the wound.

Physical Contradictions are fundamental, and can be found within each Technical Contradiction. If you don't like Technical Contradictions, it's always possible to 'translate' a contradiction from Technical to Physical. So, for the cheese grater described in the preceding section, you could also describe it as needing to be both big and small: big to grate lots of cheese and small to control where the cheese goes.

Physical Contradictions are also easier to apply in non-technical situations. Take staffing in shops. You may want a lot of staff so that you can deal with high levels of customer demand; but you don't want lots of staff when you have fewer customers. The solution is to separate in time, and have different levels of staffing according to customer demand, for example, employ more staff in the busy holiday periods. You also don't need to translate Physical Contradictions into TRIZ language, as you do for Technical Contradictions. You can just use your normal language, but ensure that the words you use are as simple as possible to make for clear thinking and undisputed understanding of what you really need.

Physical Contradictions are so-called because they describe the physical properties of whatever system you're working with. The contradiction lies within the physical nature of your system. So when describing what you want from a tent, you could say you want accommodation that's easily portable. How do you get that? By creating something big (when you sleep in it) that becomes very small when you want to transport it. The Physical Contradiction isn't between the benefits you want but how you get them: the contradictions in the functions or features of your solution.

A system can be a product or a process; technical or management. A *system* in TRIZ-speak is just whatever is delivering your needs.

Resolving Physical Contradictions

When you've identified that you have a physical contradiction, you can resolve it by working out how you can *separate* the opposite things you want. In the tent scenario, you've identified that you want it big and small. You then have to work out how you can separate these things. In TRIZ, you can separate in four ways:

- ✔ **In Time: Having opposite things at the same time.** Do you want a tent to be big and small at the same time? No: you want it small *when* you're carrying it and big *when* you want to sleep in it. So you can separate in time; that is, you need a solution that has these opposite features at different times – it somehow changes.

- ✔ **In Space: Wanting opposite things at the same time but in different places:** In the sticking plaster example, you want sticky and not-sticky at the same time – but in different places. There's a place for it to be sticky (on your skin) and a place for it to be not-sticky (on your wound). For this kind of contradiction, you need to separate in space.

- ✔ **On Condition: Sometimes you want opposite things at the same time, in the same place – but for different features of your system.** For example, you want a window to allow light in but not wind or rain; a colander to catch pasta but also allow water through.

- ✔ **By System: Separating by system is understanding that you want opposite things at different levels, for example in the broader context and in the detail.** For example, a bicycle chain is rigid to transmit force, and flexible to go around circular gears. If you want to move a long way to get exercise but not move so you can stay indoors, you can use a treadmill (instead of you moving over the stationary ground, the ground moves beneath you).

When you've identified how you can separate the contradiction, you can use the Separation Principles in Table 3-3 to find out which of the 40 Inventive Principles have been used in the past to solve these Physical Contradictions. You then take this sub-set of Inventive Principles as triggers to generate solutions.

Table 3-3	The Separation Principles			
Inventive Principle	**Separate in Time**	**Separate in Space**	**Separate on Condition**	**Separate by System**
1. Segmentation	T	S		S
2. Taking Out		S		
3. Local Quality		S		S
4. Asymmetry		S		
5. Merging				S
6. Multi-Function				S
7. Nested Doll	T	S		
8. Counterweight				S
9. Prior Counteraction	T			
10. Prior Action	T			
11. Cushion in Advance	T			
12. Equal Potential				S
13. The Other Way Round		S		S
14. Spheres and Curves		S		
15. Dynamism	T			
16. Partial or Excessive Action	T			
17. Another Dimension		S		
18. Mechanical Vibration	T			
19. Periodic Action	T			
20. Continuous Useful Action	T			
21. Rushing Through	T			
22. Blessing in Disguise				S
23. Feedback				S
24. Intermediary	T	S		
25. Self-Service				S
26. Copying	T	S		
27. Cheap Short-Living Objects	T			S
28. Replace Mechanical System			C	

(continued)

Table 3-3 *(continued)*

Inventive Principle	Separate in Time	Separate in Space	Separate on Condition	Separate by System
29. Pneumatics and Hydraulics	T		C	
30. Flexible Membranes and Thin Films		S		
31. Porous Materials			C	
32. Colour Change			C	
33. Uniform Material				S
34. Discarding and Recovering	T			
35. Parameter Change			C	
36. Phase Changes			C	
37. Thermal Expansion	T			
38. Boosted Interactions			C	
39. Inert Atmosphere			C	
40. Composite Materials		S		S

Clever Tricks to Outsmart Contradictions: Using the 40 Inventive Principles

Fundamentally, the 40 Inventive Principles are the 40 ways in which human beings solve contradictions. They're *inventive*: they're clever solutions that were used to solve difficult problems. But because they're also *principles* they're broad and conceptual, so you need to work out how to put them into practice.

Applying the 40 Inventive Principles to real problems

'If I were given one hour to save the planet, I would spend 59 minutes defining the problem and one minute resolving it,' said Albert Einstein, and identifying your contradictions first will help you define the problem correctly.

Generating clever solutions to the wrong problem takes you further away from where you want to be – not closer! – so devoting some time and energy to understanding your contradictions beforehand is absolutely crucial. A well-defined contradiction suggests a number of Inventive Principles, which you then apply to generate solutions.

When applying the 40 Inventive Principles, you still have to work hard. You have to apply your domain knowledge and experience: real creativity is about coming up with an idea that's not only novel but also useful. To be useful it has to be practical, which is why experience and knowledge of how stuff works is absolutely critical. Sometimes, however, your experience of what works and what doesn't (and what's been tried before and failed) can get in the way of new ideas. The 40 Inventive Principles help stimulate your creativity and hence come up with new ideas, allowing you to think freely and apply all the rules of brainstorming as you generate new ideas.

Many creativity techniques apply golden rules for idea generation (see the nearby sidebar, 'The rules of brainstorming'), and they're fantastically useful. However, they're not enough. Unfocused brainstorming for new ideas is a completely random process: you may come up with some genius solutions or you may not. Generating solutions is a bit like digging for buried treasure, and TRIZ provides the map: the 40 Inventive Principles suggest the parts of your treasure island where people have found the most gold in the past.

So apply all the rules of brainstorming but also use the 40 Inventive Principles to focus your energy and creativity in the areas where you're most likely to find inventive and clever solutions.

Learning the 40 Inventive Principles

The 40 Inventive Principles are worth learning, because they enable you to move fast when you're problem solving. It's not necessary to learn the numbers – unless you want to impress other TRIZniks at parties – but knowing what Parameter Change means, and the fact it has six different flavours, will keep your brain firing at full speed when you're solving contradictions. (Remember – Appendix A has the full list and explanations of the Principles.)

First thing in the morning, before you get stuck into emails (perhaps with your first cup of coffee), take a Principle and think about where you've seen it put into practice, and come up with your own examples. In 40 days you'll have learned them all!

The rules of brainstorming

Alex Osborn, an advertising executive, developed techniques for generating creative ideas, and he introduced the concept of brainstorming for encouraging creative thinking in his 1953 book, *Applied Imagination* (Charles Scribner). Osborn's 'golden rules' for brainstorming are:

✔ Focus on quantity not quality

✔ Withhold criticism

✔ Welcome unusual ideas

✔ Combine and improve ideas

Osborn offered this nugget of wisdom: 'It is easier to tone down a wild idea than to think up a new one.'

Digging for buried treasure in the right places

Defining your contradiction and the Inventive Principles most useful for solving your problem ensures you focus your energy on the most useful solutions for you. You step through the Prism of TRIZ and apply those solutions that other people have found particularly useful for this kind of problem. This gives you real confidence, and if you get a Principle you think can't be useful, still stick with it! Most people reject Inventive Principle 32, Colour Change, when it's first presented, but it can form the basis of some clever and practical inventions. It's often the Principle that initially seems the most unlikely that provides the breakthrough – for the very reason that no one would intuitively look there.

Imagine you want a dressing for a wound that protects it from dirt and germs but allows you to track how well it's healing. This is a Physical Contradiction that you can separate on condition, and one suggested Inventive Principle for doing so (explained in the 'Resolving Physical Contradictions' section earlier in this chapter) is 32, Colour Change, which would suggest a transparent dressing. Figure 3-8 shows the Prism of TRIZ to find this solution.

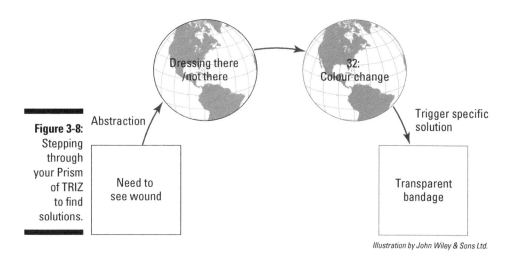

Illustration by John Wiley & Sons Ltd.

Figure 3-8: Stepping through your Prism of TRIZ to find solutions.

Applying the 40 Inventive Principles helps you find clever solutions: you're prompted to apply your knowledge in new ways, and look for solutions in new places. Sometimes you may not know how to do something – make transparent wound dressings, for example – but TRIZ prompts you to go and find out, directing you in a focused way to the places you need to investigate to locate the knowledge you need. So as your experience and knowledge grow, so does your ability to apply the Principles in pragmatic ways: your ability to use TRIZ effectively increases, directing both your expertise and your creativity to where you're most likely to find clever solutions.

Chapter 4

Applying the Trends of Technical Evolution

*T*he Trends of Technical Evolution are one of the most exciting and easy-to-use tools in TRIZ. Anyone who has to develop new products or services will find them both incredibly powerful and useful, as they help predict the future of technical systems.

The Trends are a broad tool, offering you directions and suggestions of routes forward; you need to use your domain knowledge, expertise, skill and creativity to turn the conceptual solutions of the Trends into practical new solutions, but they're the most focused and successful method for helping you lift your head and look to the future.

The *Trends of Technical Evolution* are one of the TRIZ tools based on extensive research and patent analysis. As soon as the 40 Principles (outlined in Chapter 3) had been uncovered, the TRIZ community turned their attention to the development of systems. What they observed was that technical systems tend to evolve following the same certain patterns; that is, systems in very different applications and industries nonetheless follow similar stages of development. Like the 40 Principles, these patterns were uncovered, not invented, from across many different technologies and industries.

The Trends are analogous to the evolution of biological systems: after a system has been 'born' or invented and launched on the market, it must adapt according to environmental conditions in order to survive. Market and customer needs change and develop, and the Trends demonstrate to you the patterns systems can follow to evolve to continue to better meet these changing requirements and respond to competing products. The Trends only apply to systems that have been launched successfully, thus showing that they have met a need and delivered benefits (see the nearby sidebar, 'Systems meet needs').

The Trends help you to start with a successful area and develop new systems that may offer very innovative and different ways to meet that fundamental need; the market you're targeting, however, already exists. This removes some of the risk from innovation, and provides you with structure and prompts to find new solutions.

Systems meet needs

All problem solving is a matter of developing systems that provide benefits. Invention is the creation of a new system that's designed to meet previously unmet needs. A system can be anything: a physical product, technology, process or business model, for example. You can even create an integrated system by combining a number of these things. The Trends help you continue to develop your systems over time to help them better meet needs. The environment (market conditions, competitors, customer needs) constantly changes, and, while the results of market research can be useful, they can't provide the whole picture, because the changing environment can also be driven by the systems themselves. New technologies can open up new possibilities and create new customer needs, and great inventions sometimes fail for reasons that are hard for organisations to predict.

Systems only develop in order to meet changing needs, but instead of attempting to capture and define all potential needs and how they evolve over time, the Trends focus their attention on the systems themselves, using them as the starting point for predicting the future. The Trends show you the patterns other, analogous systems have followed in order to achieve and maintain market success – so that you can follow them with your own systems.

The Trends have captured how successful systems have evolved and developed in the past, distilling the patterns of success. When the next generation of a system is launched – a new product, process or service offering – it often (but not always) heralds the beginning of the end of the old system. The Trends allow you to be the one who releases that next generation ahead of your competitors.

Looking More Closely at the Trends

In total, 8 Trends exist, as shown in Figures 4-1a through 4-1e. As with the other solution tools (the 40 Inventive Principles, the Effects Database and the Standard Solutions), the Trends are conceptual in nature: they strip away the detail of specific industries and products to a series of more general steps or patterns.

Each of the Trends is described in this chapter in an abstract, conceptual form followed by an example. Some of the Trends have a number of Sub-Trends or lines of evolution that describe subtly different elements of them (for example, Increasing Segmentation and Use of Fields can mean the object, the surface, the space inside or the rhythms of the system become more segmented).

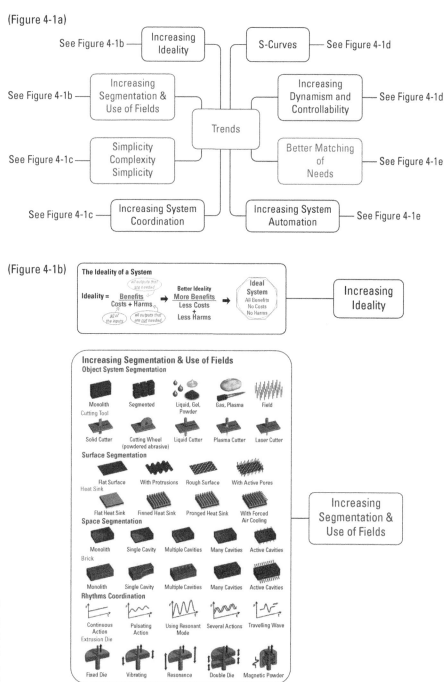

(Figure 4-1a)

(Figure 4-1b)

Figure 4-1:
The Trends
of Technical
Evolution.

(continued)

(continued)

(Figure 4-1c)

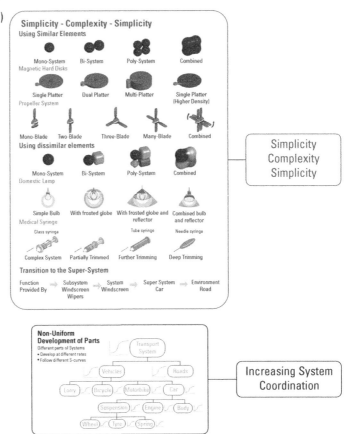

Trends aren't laws or rules that systems must follow. Rather, they're a general direction in which your system is changing and developing. Not every Trend will necessarily apply to every system – some will be more applicable than others – and not every system will follow every step; sometimes they can jump one or two. The crucial point is that Trends chart the likely future route of your system, and it's thus important to understand the general meaning of each one.

(Figure 4-1d)

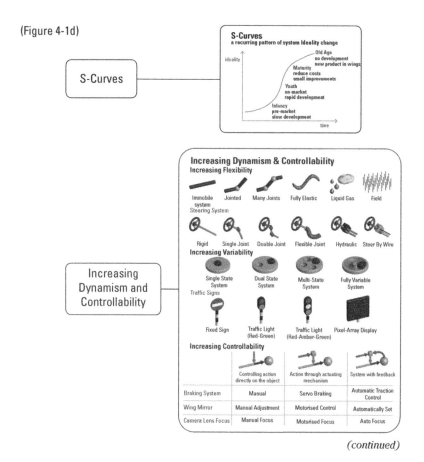

(*continued*)

Discovering the 8 Trends

TRIZ doesn't really accommodate feelings. TRIZ is fantastically logical and the Trends predict the future directions for your systems. The Trends don't tell you why or how or when – they just say that eventually this will happen. It's up to you (and market research) to work out the timings, and to develop the technology to make it happen. What the Trends do is predict the direction of your technological development, giving you practical prompts and suggestions of how you can make that future a reality.

(continued)

(Figure 4-1e)

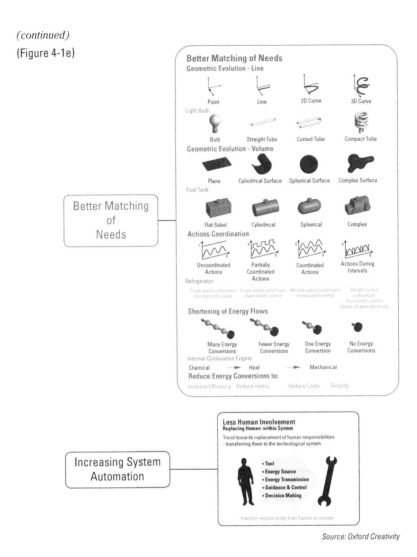

Source: Oxford Creativity

As soon as you become familiar with the Trends, you start to see them every-where: they're simple descriptions of how systems tend to evolve from one generation to the next. The Trends don't have a ranking or order of impor-tance, and you don't need to apply them in any particular order. The only exception is Increasing Ideality, which is the most fundamental of all Trends and states that systems get better over time.

Let's spell them out now – here are the 8 Trends:

1. **Increasing Ideality:** Systems improve over time by providing more benefits with fewer inputs/costs and resulting in fewer harms.

2. **S-Curves:** Systems improvement follows a typical pattern over time; that is, initial slow development after invention, rapid improvement in youth and on the market and slower rates of improvement in maturity.

3. **Increasing System Coordination** (or Non-Uniform Development of Parts): System components become more coordinated over time; different parts of the system develop at different rates.

4. **Increasing System Automation** (or Less Human Involvement): Systems become more automated and self-sufficient.

5. **Increasing Segmentation and Use of Fields:** Systems become segmented by being broken into ever-smaller pieces, until they become a field effect.

6. **Increasing Dynamism and Controllability:** System components work together more dynamically by becoming more flexible and variable. As a result, they need more control.

7. **Simplicity–Complexity–Simplicity:** Systems follow a cyclical Trend, whereby they start simple, become more complex as new elements are added and then become simple again as a result of integrating those new components.

8. **Better Matching of Needs** (or Matching and Mismatching): Needs develop over time, and systems become better matched to all needs and become more efficient.

The Trends are very easy to understand when you see examples, and the pictures help, so I'll run through a few examples of the Trends in action with the help of some pictures.

One of the Sub-Trends of Increasing Dynamism and Controllability – Increasing Flexibility – says that things start rigid and become more flexible as a result of becoming hinged, then the next generation becomes elastic, until they finally turn into a field effect (that is, some kind of field, for example, laser, magnetism, ultrasonic and so on). One example is the evolution of measuring systems, as shown in Figure 4-2.

The Trend of Increasing Segmentation and Use of Fields simply means that things tend to get smaller as a result of being segmented, until they become so small they're field effects. One example is detergent for cleaning clothes: initially a bar of soap, it then became soap flakes, followed by powders, liquids and gels. The first field-effect washing machine has now entered the domestic market; it uses ultrasound to clean clothes, as shown in Figure 4-3.

Immobile system · Jointed · Many joints · Fully elastic · Liquid gas · Field

Measuring devices

Ruler · Hinged ruler · Many hinged ruler · Tape measure · Laser measure

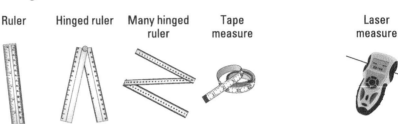

Illustration by John Wiley & Sons Ltd.

The Trends are very logical and predict the next steps for any kind of technical system. What the Trends don't tell you is *when* these next steps will be necessary. This is usually because the drivers and inhibitors influencing the system aren't within the system itself; rather, they're either outside it in the bigger picture or in the details of the technology delivering the system.

Monolith · Segmented · Liquid, gel or powder · Gas, plasma · Field

Methods to clean clothes

Bar of soap · Soap flakes · Powder, liquid, gel detergent · Ultra-sonic cleaning of clothes

Illustration by John Wiley & Sons Ltd.

For example, one of the Trends predicts that systems will require less human involvement. Responsibility for things happening moves from the human operating the system to the system itself, following a series of steps, as shown in Figure 4-4.

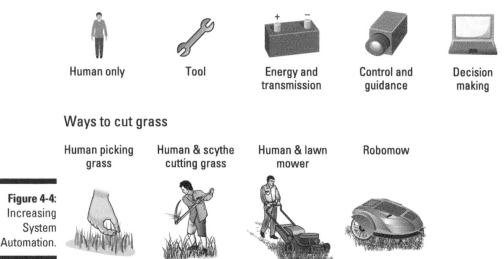

| Human only | Tool | Energy and transmission | Control and guidance | Decision making |

Ways to cut grass

| Human picking grass | Human & scythe cutting grass | Human & lawn mower | Robomow |

Figure 4-4:
Increasing
System
Automation.

Illustration by John Wiley & Sons Ltd.

As Sister Sledge almost sang, 'Systems are doing it for themselves'. Less human involvement and more automation is increasingly seen in all kinds of systems. For example, many systems inside a typical car now require less human interaction – from your automatic windscreen wipers that match the weather to automatic braking, to intelligent wheels that adjust according to driving conditions. In fact, the technology exists to create driverless cars. What holds it back are factors in the environment such as public opinion and a feeling that humans need to retain control of this Trend's final step, decision-making. The technology that would make it possible for planes to fly completely autonomously has also existed for some time but is resisted through a sense of mistrust and fear. People's concerns are what hold back planes from taking this last step in the Trend, not available technology.

Charting future product directions

One of the most effective uses of the Trends is charting the future of your product. The Trends are particularly powerful when applied to real, practical systems. They can be used to develop ideas further, but because they also describe how real technical systems evolve, they're especially useful when you want to develop an existing technical system.

The Trends are easiest to use when you know not only the present incarnation of your product but also its past: the Trends work by charting the future from the past and the present. They form a trajectory of technology, and the more information you have on previous generations of your system's past, the better able you are to map and look further into the future.

The Trends provide you with conceptual suggestions regarding where your next-generation system may go in the future. The prompts themselves can seem rather abstract, but when you apply them to real-world, existing products it's often very easy to see the many different ways in which these could be put into practice to generate new, practical systems.

Consider mobile phones. The first models were without doubt imperfect. They were very expensive, big and heavy, largely because the battery was so huge and didn't work very well. However, they had one big benefit that other systems didn't deliver: the ability to make a phone call from any location. One of the things the Trends tell us is that even though a product may not be perfect when it's launched, it will improve. The most fundamental of the Trends is Increasing Ideality: once a system has launched it will get better, and you know how – by increasing benefits, decreasing costs and decreasing harms, as shown in Figure 4-5.

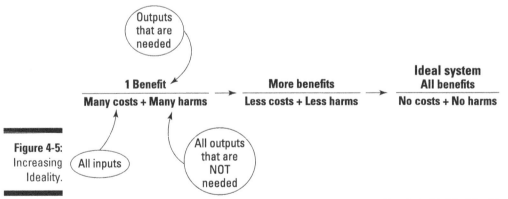

Figure 4-5: Increasing Ideality.

Illustration by John Wiley & Sons Ltd.

When a new product is first developed, it has one clever new function that meets a previously unmet need. A phone that you can use on the go is one example; another is an electric bicycle. Its clever new function is adding power, which makes riding the bike easier for people who may lack the stamina to cycle up hills and also enables longer distances to be covered. However, when any new system is invented, it's pretty rubbish in almost every other respect. It's probably very expensive, has lots of problems and may not even deliver its main function terribly well.

The Trends tell you not only how your system will develop and improve, but also, crucially, its rate of change over time, which will follow an S-Curve, as shown in Figure 4-6. We can understand the development according to how it's Ideality changes over time.

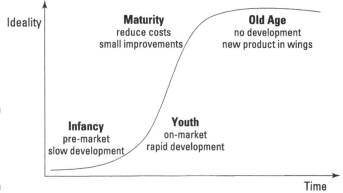

Figure 4-6:
Under-
standing
S-Curves.

Illustration by John Wiley & Sons Ltd.

Development isn't linear. Each product develops in four main stages, which have different characteristics:

- ✔ **Infancy:** Your product is in development and not yet on the market. At this point you're pumping money in. The system has lots of problems that still need to be solved. You have one main benefit, but lots of costs and lots of harms, so the system's Ideality is very low. When you've reduced the costs and harms to the point at which the benefits out-weigh them and you've achieved a balanced Ideality, you can release the product onto the market.

- ✔ **Youth:** The product has hit the market and is improving rapidly. You start to make money and solve problems: big contradictions are resolved and you work out how to make the product and its components better, cheaper and faster as you hit economies of scale. Costs and harms are reducing, but, more importantly, benefits are increasing as you add new, exciting functions.

- ✔ **Maturity:** Increases in Ideality slow down as the product and its market mature. You can add very few new benefits, so the focus is mostly on cost reduction and incremental improvements.

✔ **Old age:** The system is fully mature and Ideality flattens out or, in some cases, starts to decline. Costs and harms continue to be removed. Actually, at this stage in the product's lifecycle, even benefits are removed if this action can reduce costs and harms more. You prepare for the end of your product's life.

So, what happens at the end of an S-Curve? A new product is waiting in the wings, with a new S-Curve! Crucially, when it's first launched, this new product usually has a lower Ideality than the mature product it's replacing (see Figure 4-7).

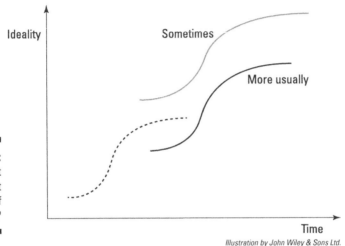

Figure 4-7:
What happens at the end of an S-Curve?

Illustration by John Wiley & Sons Ltd.

This situation arises because, while the new product has a new benefit, it's also still in its infancy or youth and thus needs further development in order to sort out its associated costs and harms. Its lower Ideality than the existing product makes it seem unappealing in comparison; however, because the new product is at such an early stage in its lifecycle, it's capable of much larger improvements and will eventually overtake the mature product. Some exceptions to this rule exist – memory sticks, for example, offered so many more benefits than floppy disks (large memory, robust, easily transportable) that they quickly superceded them. More usually, though, a number of new S-Curves are evident that are currently worse than those of your existing product. Understanding how systems evolve following S-Curves allows you to see the importance of choosing, investing in and developing new technologies. If you aren't investing in the next generation of your systems, someone else will be.

The modern definition of a Kodak moment

Kodak was so famous that one of its advertising slogans – a 'Kodak moment' – became a well-known phrase to describe an event that was worth capturing in a photo. Kodak held one of the first patents for digital photographic equipment, but was so heavily invested in the production of photographic film, it failed to see how it could change the market. Digital cameras not only change the way we take pictures but also remove the need to print photos at all: as the environment surrounding a camera (your computer, the Internet and so on) has developed, so it has become possible to store, share and view photos electronically. The modern definition of a 'Kodak moment' is now a business that fails to look to the future (if only it had used TRIZ!).

Understanding that film photography was at the top of its S-Curve might have given Kodak the motivation to explore alternative technologies: mapping out potential future scenarios in 9 Boxes (covered in Chapter 8) would also have highlighted how the environment could provide new opportunities and the exciting future directions in which photography was headed.

When digital cameras were first launched, they weren't as good as film cameras. They were more expensive, bigger, heavier and didn't take such good photographs. However, they offered many more benefits than film cameras, and as the technology (and infrastructure) was developed and improved, digital cameras outstripped film cameras – and wiped out their market. Kodak failed to make this shift, as explained in the nearby sidebar.

If you have a fully mature system, all may not be lost, however. For a start, some systems sit at the top of their S-Curves for a very long time: the same sand filtration technique for cleaning water is still used in many swimming pool filters, and it was invented by the Ancient Romans! Also, a niche market may well exist that needs a key function. In the case of film cameras, many photographers still appreciate film, and Polaroid cameras are used by many police forces as they have one benefit that digital photos lack: they can't be changed.

Achieving a system's ultimate destiny

The ultimate destiny of any system is to disappear. This TRIZzy phrase appears a little 'out there' but is nonetheless wise. What it means is that any system will disappear because:

- ✔ The needs met by the system will have disappeared.
- ✔ The benefits delivered by the system will be delivered by the super-system.

An example of a system disappearing is vaccination rather than treatment of an illness: if you can develop immunity to an illness, you no longer require the benefits of treatment. In the case of smallpox, for example, vaccination was so successful that the disease was completely eradicated; vaccination was no longer necessary because the need it met (immunity) disappeared. Digital memory is another example: technology for fast streaming means that you can now hold your data, photographs and even computer programs in cloud storage (where data are stored remotely, across multiple servers, and managed by a hosting company – and accessed only when needed) rather than paying for ever-larger data storage. This Trend is also currently at work with music, films and TV shows: with the advent of subscription services, people are realising they can get the benefit of easy, lower-cost access to the content they want without incurring the additional cost in space of storing physical CDs and DVDs (and their players) in their homes.

This approach isn't just a warning but also a philosophy. The most fundamental of the Trends is Increasing Ideality, which states that systems get better throughout their life until they achieve perfect Ideality; that is, all benefits and no costs or harms.

The ultimate destiny of all systems is that somehow you get everything you want, with zero inputs and zero harms. You can achieve this outcome by either no longer needing what a system delivered or getting the benefit elsewhere. You must be conscious that whatever you do will one day disappear. Countless stories exist of businesses not seeing the future, or of being so focused on the details of their products that they've lost sight of the changing world around them, and their customers' needs. Thinking in Time and Scale (Chapter 8) in conjunction with the Trends will help you look to the future, but having Ideality at the heart of how you understand your products will help keep your customers' needs at the centre of your thinking.

By focusing on Benefits, you focus on what really matters – the outcomes your customers want. Systems exist to meet needs: if your customers discover there are other, better ways of meeting those needs, you will lose them. Keeping those needs at the heart of your product strategy – and applying the Trends – will enable you to create and develop the right products to meet those needs.

Applying the Trends

Applying the Trends systematically enables you to develop the next generation of your products and services.

The Trends enable you to look at your system with new eyes: you learn to strip out unnecessary detail and consider your system in a more abstract, conceptual way. This alone helps you consider new possibilities, and then when you look at the Trends, they'll suggest conceptual new solutions or developments for your system.

Listening to the voice of the product

One of the lovely things about the Trends is that they start with the product – not the market. The DNA of the product itself will tell you how it's likely to develop. Yes, you need to listen to the voice of your customers and study your market, but you should also listen to the voice of the product. The Trends help you analyse the different elements of your product and will suggest in which directions your product is likely to develop.

To listen to your product you must have a clear understanding of its history, industry and how it works. Your domain knowledge is critical here: if you have a lot of experience in your industry, the necessary knowledge is at your fingertips. The Trends will help you utilise this knowledge most effectively to generate new ideas and new products; they're prompts or triggers to suggest what the future of your system will look like.

One of the most useful Trends for breaking psychological inertia regarding your system is Increasing System Coordination. This Trend states that different parts of your system will develop at different rates. Over-developing those parts of your system with which you feel most comfortable and understand best is a typical approach. Elements of your system, however, will be relatively undeveloped, and these will be holding it back. Identifying the least-developed parts of your system – those lower down on the S-Curve of their development – and focusing on these will result in dramatic improvements in your system. Listening to the voice of your product is important because it challenges your thinking about your system. Too often you become focused on the details of your system or the parts of the technology that you understand best. This Trend helps you re-evaluate your system and identify where you need to be focusing your attention.

The Simplicity–Complexity–Simplicity Trend states that systems start simple, become more complex by adding new components (either duplicating existing components or adding new ones) and then become simple again as a result of integrating all the components into a combined system. Figure 4-8 shows this concept graphically using the example of a lightbulb.

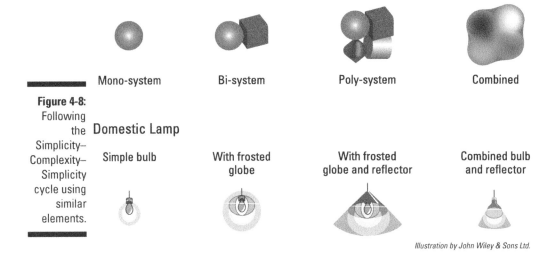

Illustration by John Wiley & Sons Ltd.

Figure 4-8: Following the Simplicity–Complexity–Simplicity cycle using similar elements.

To apply this Trend you need to identify parts of your system that have added another feature. You then consider a number of other features you can add, and how you can bring all these together into a single combined and integrated system. Consider, for example, the added features on your smartphone (alarm clock, pedometer, diary, notebook, camera and so on); razors, too, have added similar components (a single blade to two, three, four and even five blades) and dissimilar components (moisturising strips, shaving foam). Many shops apply this Trend: department stores offer many different concessions; petrol stations supply coffee, food and cash points.

Using the Trends to develop a next-generation system

The Trends are particularly useful for developing the next generation of any kind of product.

The steps are simple:

1. **Pick a trend.**

2. **Identify where on the Trend your current system lies.**

3. **Use the next step on the Trend as a trigger to suggest new ideas or new products.**

4. **Repeat this for every step on the Trend – up to the very end.**

5. **Pick a new Trend and start again!**

Apply the Trends at different levels of scale (Chapter 8 deals with scale). This means thinking not only of the product itself but also the big picture – the super-system – and the detail – the sub-system or components.

Considering how the super-system surrounding your product will change helps you also think about new ways in which your system could interact with the environment, which may suggest new opportunities.

Consider the smartphone. You could apply the Trends to the system surrounding the device; for example, changes in networks or regulations that affect networks. Will radio frequency ranges change, or will Wi-Fi availability change, and what impact may that have on your device and how people use it? How about your customers: will their needs change, where in the world will they be located, will you attract new customers (think about more children using smartphones and also an increasing elderly population)? Even though you may not be responsible for, or able to control, these changes in the super-system, the Trends will nonetheless predict likely future directions, which will provide opportunities for you to take advantage of.

You can then look at your system as a whole and consider what each of the Trends predicts for its likely development. Some examples of Trends currently in action are Increasing Flexibility: Samsung has announced the release of a flexible smartphone; and Simplicity–Complexity–Simplicity: the availability of smart watches. These Trends predict that other systems will be added that will link up with your system (for example, Google Glass) and eventually they'll all be bundled into one integrated platform.

Consider the most important components of your system, map them onto the Trends and predict how they're likely to develop. Think about processors, batteries, screens, motion sensors, microphones, speakers, and even accessories such as chargers and headphones. Again, even if the development of these components is currently outside your control, there's no reason why you can't plan for the future or even co-develop the next generation with your suppliers, using the Trends.

The Trends will provide you with many, many new ideas and solutions to current problems. When you've generated multiple ideas, group and rank them in order to select the most promising to take forward. Given the nature of the Trends, which suggest increasingly revolutionary new systems the further up the Trend you go, it can be particularly useful to chart your product's future from the ideas generated. You can plan for both small incremental changes and then a series of more revolutionary step-changes for both the near and distant future – all starting from your current product.

Moving systems forward and knowing when to go back

So, what happens at the end of a Trend? Sometimes a system has become fully mature and there's nowhere for it to go on that particular Trend. In that case, you can shift your attention up a level and look at how the Trends can apply in the environment surrounding your original system. Alternatively, other Trends may become more important; pixel-array displays are a good example of traffic signs that have become fully variable and thus reached the end of the Sub-Trend of Increasing Variability. However, the Trend of Increasing Segmentation could become more relevant: the pixels are an example of a system that's become more segmented, which suggests moving the sign to a liquid or plasma display, then to a field effect such as a holograph.

Sometimes systems can also move back along a Trend because their Ideality changes as a result of altered boundaries. As you begin to consider the broader picture, new costs and harms are revealed, which reduces the Ideality of the system. In order to increase Ideality in this situation, you can sacrifice benefits if doing so means that costs and harms can be drastically reduced.

Shifting Ideality is increasingly important as the world becomes more aware of the interaction between systems and their environment. The perceived costs of systems increase as you widen your scope: this makes you more aware of how many true costs are involved in producing your system, such as the need for power, water and materials to manufacture, store, ship and use your product – even if those inputs don't make it into the product itself. Processing cotton, for example, uses a huge amount of water. The perceived harms increase as you consider the impact of a system on the broader environment, including emission of heat and carbon, energy losses, pollution and other emissions created in the manufacture, storage, shipping and usage of the product. Disposal also needs to be considered. The changing environment can also produce new harms; for example, staff at the Kremlin have reverted to using typewriters instead of computers, in order to keep communications safe from computer hackers.

Disposable nappies are a good example of working back along the Simplicity–Complexity–Simplicity Trend. Disposable nappies are at the end of the Trend: they've integrated many of the individual elements of towelling nappies, such as replacing safety pins with sticky tape as a means of fastening and providing a robust, waterproof outer layer to replace the stiff exterior rubber, but recently many new parents are returning to cloth nappies. How could TRIZ have predicted this situation? In this instance, the Ideality of the system as a whole has changed. Consumers are recognising the harmful impact of disposable nappies on the environment, which previously had not been considered.

Systems can go backwards along a Trend when Ideality changes: new benefits, new costs and new harms are uncovered or articulated, and the system goes backwards to accommodate them. When new technologies are developed, the system will once again progress forward.

Applying the Trends More Generally

The Trends are TRIZ tools based on engineering systems, and they were designed and developed with the intention of improving technical systems. However, some Trends can apply in more general settings, as shown in the following sections.

Increasing Ideality

The most fundamental of the Trends, Increasing Ideality, demonstrates how *all* systems (technical and non-technical) can improve as a result of increasing benefits, reducing costs and reducing harms.

If you apply the Trend to how a team functions, for example, it could suggest that the team increases its output, takes less time to do so and makes fewer mistakes.

Transition to the Super-System

The Transition to the Super-System Trend states that the ultimate destiny of any system is to disappear, because either the need it addressed no longer exists or the super-system now provides the benefits it once offered.

This need not be the end of your business, however. Wilkinson Sword was able to find a new market when duelling became less popular as a means of settling disputes and is now well known as a manufacturer of razor blades. Planning and clear thinking are required, though.

Increasing System Coordination

Different parts of your system develop at different rates; some parts will be over-developed, some under-developed. This Trend suggests looking for and improving the weakest link, and also being prepared to reduce the Ideality of specific components if doing so can increase the Ideality of the system as a whole.

You can observe this principle at work in attempts to improve manufacturing flow. It can also help you understand and improve organisations as a whole. Some parts of an organisation may work fantastically well, such as the supply chain, purchasing and accounts; others, however, such as customer service, may be less well developed. This Trend shows you where to look to improve your system as a whole.

S-Curves

No one knows whether S-Curves were independently uncovered by Altshuller and his associates (the developers of TRIZ), or simply adopted from one of the many different approaches that used them, including many business schools that developed them as a way of understanding that the rate of system development isn't even over time. Most approaches plot value (function/cost) against time. I find it is essential to consider Ideality/Costs + Harms.

A person I trained showed me how he used Ideality to compare different business models. He was tasked with developing new ways of working for his company, and said that including harms was essential because it allowed him to include elements of risk and unintended outcomes (for example, reduced customer satisfaction) to make the most sensible decisions.

Using the Trends to Create Strong Patents

One of the most powerful applications of the Trends is in the development of strong intellectual property. Because they evolved from patents, Trends work spectacularly well when applied to this area. When you need to look to the future and protect potential products, let the Trends be your guide.

Strengthening and ring-fencing your own patents

If you want to excite a patent attorney, show her the Trends (whether you'll be able to tell that she's excited is another matter). Using the Trends, a clever engineer and a good patent attorney could create fantastically strong patents in an embarrassingly short period of time.

The Trends help you to take any system and predict its likely directions for improvement, and part of the work of a good patent attorney or patent engineer is to suggest modifications or alternative ways of delivering the same functions. What the Trends offer you is a systematic series of suggestions regarding exactly how you can do so. Using the Trends helps you think of new embodiments of the same idea or developments of existing embodiments.

Work through the Trends for your system as a whole and for each individual component. While this process seems laborious, in reality it's fast: the Trends are very visual, and easy for even novices to pick up and use with very little explanation. I've conducted two-hour sessions that have generated many ideas and resulted in 10–20 new patent applications.

Leapfrogging competitors' technology

If someone else comes up with the next generation of your product or launches an innovative new system, bringing out something similar isn't always the best option because you'll simply be playing catch-up. Sometimes it's best to focus instead on the generation after your competitor's technology, particularly if your organisation holds a large market share and has done so for a while. Here, your organisation may experience psychological inertia regarding how things could and should be done. It may have invested heavily in the current way of operating, which will make being nimble and investigating completely different methods of delivering benefits to its customers difficult.

Applying the Trends can be used to completely leapfrog someone else's technology. The Trends offer a very systematic and structured approach to developing new technologies, products and processes, with a successful and realistic starting point – your existing system or product.

Chapter 5

Improving Ideality by Using Resources

The main purpose of TRIZ is to increase Ideality. And this chapter is a great place to start thinking about how to do so.

Ideality is how you assess how good something is: you consider the proportion of benefits (all the things you want) over its costs (all inputs) and harms (all outputs you don't want). Measuring Ideality is a quick and simple way to understand the pros and cons of your systems and processes: you can start by checking you have all the benefits you want and also identify any problems and see how to solve them.

Your inputs (costs) result in functions, which give you two kinds of outputs: outputs you want (benefits) and outputs you don't want (harms). When you problem solve, you improve Ideality by improving your functions; that is, reducing harmful functions, improving insufficiently useful functions and reducing input functions.

By far the most elegant way of improving systems is with the wise and clever use of all your available resources, as explained in the following sections.

Understanding the Ideality Equation: How TRIZ Defines Value

Traditional methods for calculating value typically talk about either benefits over costs or functions over costs. Many approaches muddle the difference between benefits and functions, and don't consider the downsides you don't want.

The TRIZ Ideality approach to value provides a richer approach. By clearly separating benefits and functions and considering their relationship (both as it is and as it could be), you gain clear thinking and see new possibilities. By considering harms, you uncover potential problems and unintended outcomes.

Drawing all these together helps you in terms of both decision making and the development of the right systems – systems that give you enough of what you want, at an acceptable cost and with no nasty surprises.

Defining benefits

A *benefit* is something you want. It doesn't contain a description of *how* you get what you want; rather, it tells you only *what* you want – the outcomes or results your system gives you. The *prime benefit* is the one main thing you want your system to deliver.

A list of all the benefits you want forms your Ideal Outcome (see Chapter 9 for more about defining the Ideal Outcome), as it helps you understand all the things you'd have in an ideal world. Essentially, it's a wish list that helps you break out of current thinking and see new possibilities.

The first step is to clearly understand the benefits that different systems give you. Not mixing up benefits, functions and features is important.

A benefit is the outcome you want; it doesn't tell you how to get it. The function is the *how*: it delivers the benefit you want. The function is delivered by a feature in a real-life system (see Figure 5-1 for a diagram describing these relationships).

To get to the benefits you want, you continually ask, 'Why is that good? Why do I want it?'

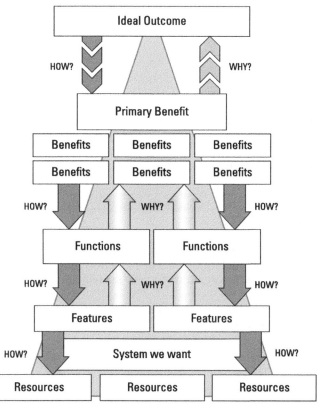

Figure 5-1:
The relationship between benefits, functions, features and resources.

When my father was buying a new car, he said it had to be big. Big isn't a benefit, I told him: big is a feature. 'Why is big good?' I asked. 'Because that states it's a managing director's car,' he replied. The benefit he was after was an impressive car, which conveyed status. This benefit may be achieved in lots of other ways that don't involve a big car (he'd look good in a Ferrari, for example!). However, my father also has many grandchildren to ferry around, so 'big' may be a useful feature that could deliver this other benefit as well.

Consider the example of buying a car and how you'd define the relevant benefits, costs and harms:

✔ Prime benefit – transport

✔ Other benefits

  • Safety

  • Personal space

- Transport other people
- Transport stuff
- Reliable
- Predictable
- Attractive
- Protects from environment
- Personal entertainment
- Enjoyable driving experience

Defining costs

In TRIZ, a cost refers to more than just money. A *cost* is any input required to deliver a function, including time, design effort, energy for manufacture, resources going into the creation of a new system and so on.

Other costs include maintenance time and energy, so for the car example you could consider costs such as:

- ✔ Cost to buy
- ✔ Insurance
- ✔ Road tax
- ✔ Service and repair costs
- ✔ Fuel
- ✔ Spare parts
- ✔ Cleaning
- ✔ Depreciation
- ✔ Time spent on maintenance (checking tyre pressure, oil and water; taking to garage)
- ✔ Parking and garage space

Defining downsides

In TRIZ, a downside is defined as a harm. A *harm* is an output from or function in your system that you don't want. The TRIZ definition is brutal: any output that isn't a benefit is a harm.

Neutral outputs or functions don't exist. Even if it isn't actively harmful, if your system produces something and you don't want it, you call it a harm. Everyday examples are obvious harms like noise and heat produced by a refrigerator and limescale deposited on a showerhead; less obvious are any features which are not obviously harmful but are there and we don't use because we don't want – such as buttons you don't use on a remote control.

Like benefits and costs, what's considered a harm may vary between users or stakeholders. Continuing our example of the car, you might consider the following harms:

- Environmental pollution from burning fuel
- Noise
- Dangerous in a crash (to driver, passengers, pedestrians, other drivers, wildlife and so on)
- Damage to other cars
- Road wear
- Speeding and parking fines
- Environmental impact at end of life

Putting the Ideality Equation together

You combine your benefits, costs and harms into an Ideality Equation.

The *Ideality Equation* is the ratio of benefits to costs and harms, as shown in Figure 5-2.

$$\text{Ideality} = \frac{\uparrow \textbf{benefits}}{\downarrow \textbf{costs} + \textbf{harms} \downarrow}$$

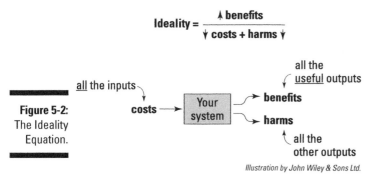

Figure 5-2:
The Ideality
Equation.

Illustration by John Wiley & Sons Ltd.

The Ideality Equation is the only one used in TRIZ, and it's not a real equation – it's a ratio! Ideality is a measure of a system's 'goodness': you balance all the benefits that your system provides against the costs you have to put in and all the harms it also generates.

When choosing between systems, rank the different Idealities by comparing the different benefits, costs and harms each system offers. You can do this in detail using numbers but a list is just as useful to help you look at the real world as it is now and to understand the Ideality of radically different systems. You can use the Ideality Equation to compare anything, for example, a house to buy or a new IT system to implement. Whenever you have to make a choice between competing options, always consider the Ideality of the different choices to help you make sensible decisions (see Chapter 9 for more on making realistic decisions).

When buying a car, it's useful to define the Ideality you want and then compare different real-world systems to identify how they measure up to what you seek. You may be able to buy a classic car for the same price as a new family saloon. These cars offer very different benefits but the former will probably require more maintenance (that is, additional costs in terms of both time and money).

Bear in mind that some benefits are universal and others specific to different people, according to their circumstances. To return to the car example, a man in his early twenties may favour something fast with a bit of panache; a father to small children may be most interested in safety features – both, however, probably value reliability.

Ideality means different things to different people. Considering the Ideality of all stakeholders is thus worthwhile to ensure that the system you choose or develop delivers the greatest overall Ideality. Don't be afraid to ask. Knowing in advance that someone may be unhappy with your system is useful, and you may even be able to do something about it!

The goal of TRIZ is to increase the Ideality of any system. You problem solve at the function level, and Figure 5-3 shows you where functions fit into Ideality.

Figure 5-3:
The goal of TRIZ is to increase Ideality.

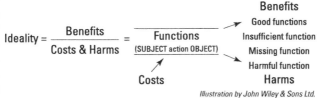

Illustration by John Wiley & Sons Ltd.

Understanding the Links Between Benefits, Functions and Solutions

A *benefit* is what you want – the outcome. A *function* is how you get that benefit. A specific *solution* delivers the function you want. For example, if you desire the benefit of clean teeth, the function that delivers it is 'remove plaque'. The commonest way to remove plaque is the specific solution of a toothbrush.

Much of TRIZ involves understanding the links between the nitty-gritty of the real world and abstract thinking. You need both. Thinking about problems in a purely abstract way may be fun but it won't be useful; similarly, if you only look at the detail of the here and now, you may miss new ways of thinking. In TRIZ you go from a specific solution to taking a step into more conceptual thinking by defining the function that solution delivers. This helps us seek other ways of delivering that function and reminds us why we have the function. It delivers benefits – an even higher level concept.

For a function to deliver a benefit it must be useful to someone. If the function doesn't deliver a benefit (no one wants it), then it's a harm.

Therefore, to be a benefit it must deliver something that someone wants, so even at the most conceptual level, benefits are connected to the real world. What's interesting is that as you take this journey between solutions, functions and benefits, at every stage you'll encounter alternative options. You can start either at the top with your benefits (described in Chapter 9) or at the bottom with solutions. You move up from solutions to functions to benefits by asking *why*. You move from benefits down to functions to solutions by asking *how*. This is the logic of the ever-popular '5-Whys' Tool.

Every time you uncover the functions your solutions deliver, an opportunity exists to find new ways to deliver those functions. When you understand your benefits, you can see other functions that can deliver the same benefits.

These are all opportunities to improve your systems and find new ways of doing things. Be open to these new possibilities and see how they can all fit together, because this thinking can offer you a route into the best use of your resources. By starting with an initial 'bad solution' you can find new, better solutions.

Say you want to buy a new airtight jar in which to store coffee (check out Figure 5-4 to see how this works). Why do you want an airtight jar? To keep air away from coffee. Why? Because coffee degases after it's ground and the flavour floats away in air. So the function you want is to maintain the flavour of ground coffee. How else can you achieve that function apart from an airtight jar? You could keep your ground coffee cold by keeping it in the fridge or in the freezer (using the resources you've got, which means you don't need to buy an airtight jar) or grind the beans just before use. Why do you want the function of flavour in your coffee? Because you want the benefit of delicious coffee. Another way in which to deliver that benefit is to move next door to a coffee shop.

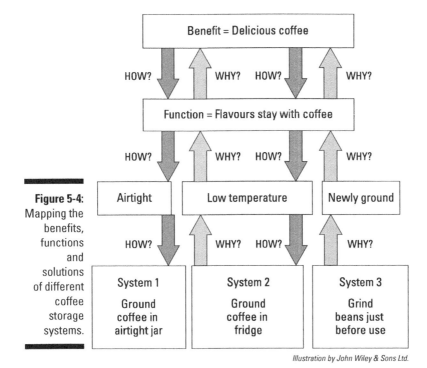

Figure 5-4: Mapping the benefits, functions and solutions of different coffee storage systems.

Illustration by John Wiley & Sons Ltd.

Understanding these links helps you understand what you want and the best ways to get it. And if you're really clever, you can look to your resources to get what you want for free.

Thinking Resourcefully

One of the most elegant TRIZ tools is also a philosophy: the intelligent use of resources to improve Ideality. If you can get something you want (a benefit) without adding anything to your system, you improve your Ideality.

Thinking resourcefully helps you build simple, elegant and innovative systems that give you what you want, at less cost and producing fewer harms to you and the environment. If you're really clever, you can take the harms around you and turn them into benefits, thus dramatically improving your Ideality. This means taking any harmful output from your system and somehow converting it into something useful for you: such as making compost out of otherwise-discarded kitchen waste.

Understanding resources and where to find them

Taking your problems and turning them into something useful is very TRIZzy and exciting. TRIZ as an approach is very much about facing up to reality and uncovering all problems rather than burying your head in the sand and hoping they go away. Uncovering all problems gives you the opportunity to do something about them. If you can actually turn your problems – any harms in your system – into something useful, your Ideality will go through the roof.

Facing up to all the problems in your system can be difficult, but those problems will still be there whether you deal with them or not. Thinking resourcefully helps you identify your problems in a positive way, in the hope that you can turn them into something useful.

Resource thinking is very much coming into its own as people are increasingly aware of the impact of human actions on the environment and the finite nature of many of the resources we take for granted. The TRIZ approach encourages us to use what we've already got rather than add anything, and also to take the harmful outputs from our systems and make them productive. TRIZ can help us make better use of our available sources of energy (such as wind, sunlight and water) and also to reduce the impact of our actions on our environment.

To use a resource, you first have to be aware that it exists. TRIZ problem solvers are aware that everything can be turned into something useful. You learn to become resourceful and inventive by initially trying to get what you want with what you already have – rather than simply throwing money at the problem. The traditional way to solve a problem is to add something; the TRIZ approach is to look around and ask whether you've already got something that can deal with the problem.

A resource is anything

- ✔ Available – in or around your system
- ✔ Already there
- ✔ You can use for free

✔ Close to the problem

✔ In your environment (sunlight, gravity, ambient temperature, pressure and so on)

✔ In the detail of your components (right down to the molecular level)

✔ Problematic, and that causes harmful things

You can use Thinking in Time and Scale (discussed in Chapter 8) to help you locate resources, because you should always be considering what's around you that can be used for free, as well as what lies in the detail of your situation.

Learning how to hunt for resources

So how do you find resources? The TRIZ approach isn't scattergun – a resource can be anything, after all . . . the shape of your system, its size, colour, weight and so on. If you list everything, you'll create an unmanageable collection of resources very quickly!

Focus on what you want. First, identify the benefits you want, then establish which functions could deliver these benefits. Look intelligently for resources when you know what you want them to do, without losing your way or ending up delivering a lot of functions that you don't really need.

This is why you often end up with intelligently simple solutions with TRIZ: you don't get distracted by all the bells and whistles of expensive and complicated solutions (which may require more costly input resources). Most systems that you buy are loaded with extra functions which you neither want nor need. How many functions do you really need word-processing software to deliver, for example? You, like everybody else, probably need only a fraction of the functions available in most software. And on top of that, all the additional functions you don't want are a harm (they can make you feel stupid, make simple actions complicated, make it hard to use and overwhelm the processing power of simpler machines) and also drive up costs (not just in terms of financial cost but also additional computer memory and space on the hard drive).

Using knowledge, experience, attitudes and emotions

When thinking resourcefully, you consider what you really want – just that – and work out how you can get it from what's already around you. This approach helps you achieve the TRIZ philosophy of getting what you want without changing anything. You find a way to achieve all of your benefits without adding anything to your system.

Often this means utilising subtle resources, such as time, knowledge, experience, attitudes and emotions. Many culture change techniques, such as Kurt Lewin's Change Model or John Kotter's 8 Step Process talk about the importance of creating the motivation to change before change can happen. Motivation can thus be a powerful resource.

You can also harness the competitive spirit of people in your team or organisation. Commonly used as a resource in sales teams, it can also inspire different ideas and solutions. The manager in charge of cleaning the facilities at Schiphol Airport in Amsterdam, for example, cleverly suggested reusing a solution he'd seen in the army (very TRIZzy!): putting targets in the urinals. Apparently men can't resist aiming for any kind of target in this situation, and this simple idea resulted in an 80 per cent reduction in 'spillage' and a 20 per cent cut in cleaning costs.

Your customers and the users of your system are an interesting resource. Supermarkets invest lots of money in researching the psychology of shopping and consumer behaviour in order to encourage people to buy more. As a result, the most expensive products are usually displayed at eye level (where people look first) and most supermarkets with an in-store bakery locate it at the back of the store because the aroma (another resource) draws shoppers towards it, past lots of other products.

Looking to the super-system and the environment

Ever-more technologies are being developed that turn readily available environmental resources into something useful. This is particularly true in the renewable energy industry, where wind, sunlight, waves and other forms of hydropower are converted into electricity. Another example of technology harnessing something in the natural environment is self-cleaning glass that mimics the surface of a lotus leaf. One kind of self-cleaning glass has a very bumpy surface at the microscopic level that encourages water droplets to form when it rains, which then roll off, taking all the dirt with them. In this way, the free resource – rain – cleans the glass.

Considering system features

Any feature of your system can be useful – its size, shape, colour, transparency, weight, roughness, density and so on – both as it currently stands and as it could be. An apocryphal tale states that the long, thin French baguette was invented by Napoleon Bonaparte. Apparently he ordered bread to be baked in this shape to make it easy for his soldiers to carry down their trousers. Although unlikely, it's a good story to illustrate this point: just as bread can be baked in any shape, so you have to remember that any aspect of your system may be useful in unexpected ways.

The U-bend outbound pipe from a domestic toilet is a good example of using the resources of shape, gravity and running water. Combined, they trap sewer gases and prevent them making your house smell.

Taking a look at sub-systems and components

Look at your system in detail to identify whether any details or components could be useful. It's particularly helpful to consider if components can be put together in new or different ways to provide useful functions or modified to deliver what you want. Bicarbonate of soda, for example, is an easily available resource in a typical kitchen, which, in addition to being used in cooking, can also clean your fridge because it changes fat into soap.

This approach is particularly important in closed systems into which you cannot bring any new resources.

Harnessing time

Forgetting that time is a resource is all too easy. How long do you have to solve the problem? Can you use idle time in your process to deliver a benefit? Examples abound in the confectionery industry of using storage and distribution time to change chocolate from hard to runny. Do you need your solution to exist forever or can you use something that provides the function you want only at the time you want it and then disappears?

Utilising space

You don't always need everything everywhere. Sometimes you can use resources from one place somewhere else, or provide the functions you need only where they're needed and nowhere else. Help may be available elsewhere that you can easily access if needed, which means that you don't need all knowledge and experience to hand or on-site.

Many shops are appropriating the supermarkets' clever use of space and displaying special offers and tempting treats at the counter. Consider, for example, the display racks containing magazines, chocolates and other low-cost items that form a barrier you have to walk past in order to reach the counter and pay.

Filling voids

Don't forget things that aren't there! Empty spaces can also be a useful resource. For example, an empty space filled with air can provide functions such as insulation or cushioning. Gaps in a fence allow air through while also maintaining privacy.

Transforming harmful outputs and waste

Utilising this resource is an excellent way to turn a harm into a benefit, and a powerful approach for all organisations looking to improve their sustainability. Biofuel converts domestic food waste into energy and a cork factory in Portugal takes the cork chippings and burns them to create energy on-site. Identify any harmful outputs and see if they can be a benefit for you – or someone else.

Getting what you want using what you have

When you solve a problem with what you already have, you create an inventive solution. Doing so is particularly useful in industries such as food and pharmaceuticals, which are subject to stringent regulations regarding what can be used in or added to products. If you can solve a problem using what you already have, you may avoid the need to go through a lengthy re-testing and validation exercise.

This approach is also useful for closed systems, whereby you can only use what's available. I often use my resource thinking when I'm in a foreign hotel room, late at night, and without a crucial item: I have to invent something on the spot. You too can let your brain run free. Identify your first top-of-the-head solutions and any resources that occur to you, then:

1. **Identify the benefit(s) you want.**

2. **Identify the function(s) that can deliver the benefit(s).**

3. **Go through the resource prompts and list all resources.**

4. **Identify which resources can deliver the functions you want.**

Ellen Domb, a great TRIZ teacher and a founding editor of the *TRIZ Journal*, created the best resource-thinking activity I've ever encountered for teaching TRIZ resource use – solve the problem of saving the lives of everyone on-board the *Titanic*:

- **Problem context:** The *Titanic* hit an iceberg on 15 April 1912, carrying 2,224 people and lifeboats for just 1,178 (meeting safety regulations at that time). The rescue ship was four hours away, and the ship was going to sink in two hours (and they knew it – the chief naval architect was on-board). How can you solve this problem using only the resources available to the passengers and crew? First, identify what you really want (see the next bullet).

- **Benefit you want:** Everybody survives! How do you get this? People die in cold water within four minutes, and drowning is also a risk. So, onto the next bullet . . .

✔ **Functions you want:** Stay afloat; stay warm. Next, identify all relevant resources, using a resource checklist as a prompt, as shown next.

✔ **Resources:**

- Knowledge, experience, feelings: Chief naval architect, engineers, fear, panic, disciplined crew, passengers' experience

- Super-system and environment: Good weather (still water, no wind), speed of boat, other rescue boat

- System features: *Titanic*

- Sub-system and component resources: Lifeboats, buoyancy aids, deckchairs, musical instruments, bedding, cooking equipment, bottles, baths, luggage, unflooded bulkheads, working engines, heating, ventilation system, winches, cranes

- Time: Two hours before sinking; only need a solution to last two more hours before help arrives

- Space: Crew of rescue ship available via telegraph

- Harms and waste: Iceberg

How can you put these resources together to generate practical solutions? Obviously, many solutions exist but some of my favourites are:

✔ Get the crew to organise the passengers to make more flotation devices in the two hours available by putting together various buoyant component resources.

✔ Make holes in the bulkheads at the other side of the ship so that it sinks evenly, instead of breaking in two. This strategy will probably buy enough time for everyone to survive until the rescue ship arrives.

✔ Move everyone onto the iceberg.

✔ Overload the lifeboats. They were designed for high winds and rough seas and the weather was calm. Possibly everyone could have fit in and survived.

Embedding resource thinking into everyday life

Start embedding TRIZzy resource thinking into your everyday life – from finding solutions to urgent problems (from 'I've forgotten my toothbrush' to 'we have declining sales') to improving systems and situations. Start by thinking that you're going to get what you want with what you already have and that constraint will encourage really inventive thinking and fuel innovation.

Starting with what you want and looking for just that – and no more – will encourage you to embrace elegant solutions rather than accept lots of functions that you don't need. Considering your Ideality will help you see that sometimes a simple system with minimal inputs that gives you a small number of useful functions is better than a complex system which offers many more functions but requires lots of costly inputs and produces many harms.

Resource thinking should become a reflex. You need to regularly ask yourself, 'How can I get what I want using what I've already got?' – searching for resources that are readily available and looking to turn any harms into benefits. In this way, TRIZ will help you make the best use of the world's resources – including your brainpower!

Chapter 6

Using the TRIZ Effects Database

. . . .

In This Chapter

▶ Finding innovative solutions with existing knowledge

▶ Using the Effects Database

▶ Inventing with TRIZ

. . . .

*W*hatever kind of problem solving you're doing – inventing or looking for solutions to immediate problems – you are not on your own. You have the knowledge and experience of the world at your fingertips: by learning to think conceptually, you can discover and reapply the relevant existing solutions in novel applications. Anyone can learn to be an inventor, and TRIZ shows you how.

When you're problem solving, using any of the TRIZ tools, you distil your real-world, detailed, factual problem to a conceptual problem. You can then see how your problem is similar to other problems the world has faced before. You can access the clever solutions other people have come up with in the past and reapply them in new ways to develop practical and innovative solutions to your problems. This chapter looks at some of the tools available to us all.

The *TRIZ Effects Database* is a unique resource within TRIZ: a catalogue of all the known scientific and engineering effects discovered so far, arranged into simple questions and answers. I explore how it can help you in this chapter.

More fundamental than the database, though, is learning that someone else may have the answer to a question or problem that you're grappling with. This chapter also shows that you just need to learn how to look intelligently and you'll find the solutions you need, whether you're problem solving or inventing.

Thinking Innovatively with the Prism of TRIZ

Thinking innovatively means reapplying existing solutions in new applications: you use existing knowledge but in new ways. To find these innovative solutions you need to learn how to look and think in the most productive way, taking a step back from the real world into a more abstract, conceptual way of thinking, without getting stuck in the detail.

Taking out unnecessary detail

The first step to thinking innovatively is to learn to strip out any unnecessary detail or technical jargon. This will feel hard at first but is a good discipline to master. Although it's not the way you're taught to speak, particularly when dealing with something within your own area of expertise, it allows for clear communication as well as clear thinking.

When you work within a specific discipline, one of the first things you learn is the language. Every field uses words and acronyms that have very specific and precise meanings and these allow you to communicate efficiently with other people within that field. TRIZ is a great example – it's much quicker than saying Teoriya Resheniya Izobretatelskikh Zadatch every time!

Using the language of your discipline makes you feel comfortable and helps you to express yourself efficiently and precisely to your peers: this is why organisations develop their own shorthand, jargon and acronyms. Your language can give you a sense of belonging but, on the flip side, can resemble a club where outsiders are unable to contribute and feel unwelcome.

As a result, you can end up working in a silo of only people who share your language, which cuts you off from the brainpower and resources of others when problem solving. If you can learn to communicate in problem solving using very simple language, you can broaden the field of your enquiry and the range of solutions on offer to you. Often the greatest innovations come from the application of knowledge outside of your domain: communicating without technical terms helps you to access this new knowledge and put it to use.

Communicating with people outside your discipline or organisation becomes hard and is prone to confusion and miscommunication when you're using your own company's jargon and acronyms – examples specific to a discipline or company can easily be misinterpreted. I once participated in a very confusing conversation with someone concerning patents; when he spoke of increasing his ROI, I thought he meant Return on Investment – in fact, he was referring to Record of Invention (his company's term for the formal submission of ideas).

Get into the habit of asking everyone to explain what their acronyms stand for. Often people can't remember or will have made them up. You certainly won't be the only person in the room who isn't sure what they mean.

Using simple language can take courage because it's very easy to hide behind technical jargon, especially when you don't really understand the situation. If you can explain something using very simple language, you know you really understand it. Some of our greatest thinkers demonstrate this fact, such as Richard Feynman (see the nearby sidebar about this chap).

Using simple language really helps take out unnecessary detail and see the wood for the trees. It also helps break *psychological inertia*, which is what happens when you get stuck in a way of thinking.

The language you use can create psychological inertia about your situation, about what you want and even how you can use what you have. Using simple, general language will break you out of that psychological inertia, allowing you to think more conceptually and open your mind to new possibilities.

Richard Feynman's simple language

Richard Feynman (1918–88) was a great theoretical physicist who not only pushed the borders of science but was also a passionate believer in explaining things as simply as possible. He described atomic theory as 'things are made of atoms – little particles that move around in perpetual motion, attracting each other when they are a little distance apart, but repelling upon being squeezed into one another'. This is a great example of explaining a really complicated theory in a very simple way that anyone can grasp, not just theoretical physicists. If we all take the same approach to explaining our problems, we'll be able to communicate much more clearly with everyone.

Explaining things in simple language is not always easy: simple expression can in fact be quite hard because you are communicating only what is essential – no more, no less. True simplicity of expression is only possible when you really understand what you are talking about; as Feynman also said, 'If you can't explain it to a 6-year-old, you don't really understand it yourself.' Using TRIZ helps you strip down your problems to uncover their core, and enables you to describe and explain them using very simple language – ensuring that everyone (yourself included) really understands what's going on.

Your specific language is often couched in terms of a particular solution or view of the problem, which will be limited by your knowledge and experience. You should be wary about the language you use, and aim for it to be as general as possible. Using simple yet accurate language gives you a broader framework within which to search for the right – and potentially very innovative – solutions to your problems.

Reapplying clever solutions in new and exciting ways

Using simple language and thinking in a more abstract way isn't instinctive behaviour for most of us. You can, however, make it a habit. Although this approach appears to take extra time and effort, it actually results in a much more efficient way to tackle your problem.

Following your Prism of TRIZ (first described in Chapter 2 and in more depth below) means that when you start looking for information and answers, you narrow your focus and attention on a smaller amount of information and the solution that will work for you. Ironically, TRIZ was developed at a time when accessing information was very difficult; it meant finding the right book or paper in a physical location such as a library. Now the opposite problem exists: a huge amount of information is available via the Internet, but how do you know which is right for you? While you can use arbitrary restrictions (most people don't bother reading beyond the first ten responses to a Google search, for example), narrowing down to things that are more likely to be useful for you is a much better strategy. Using the Prism of TRIZ will help you do just that: you distil your problem down to its essence, and can see how it's similar to other problems in the past (and in other industries): you can then find analogous conceptual solutions to your problem. You then use your own knowledge and expertise to translate these conceptual solutions into practical solutions that will work in your experience. You are looking beyond your own knowledge and experience but your search is targeted and well-defined.

When you think conceptually you're able to search for analogies systematically: analogies in problem solving mean that someone has seen a similar problem in the past and found a solution to it; the similarities are at the conceptual level, the differences in the detail.

Analogous problems and solutions are captured in the TRIZ tools that you can access reliably using the Prism of TRIZ, as shown in Figure 6-1.

TRIZ thinking means that you're thinking conceptually and focusing on the right places: where other people have found conceptual solutions to your problem. Their specific problems and solutions may well be in a completely different field to your problem; however, if you can bring their solutions into your field, you reapply proven solutions in a new application – the very definition of innovation.

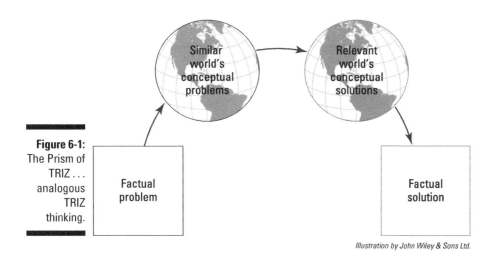

Illustration by John Wiley & Sons Ltd.

Figure 6-1:
The Prism of TRIZ . . . analogous TRIZ thinking.

Check out the two example sidebars for real-life applications of TRIZ thinking. In both of these examples, the problems and solutions are conceptually similar but the devil's in the detail. To successfully reapply someone else's solution, it's useful first to see the similarities and then to look for the differences, working out the specific needs and constraints of your own situation in order to see what can be usefully transferred as it is and what needs to be modified. Locating and using known relevant solutions puts you on the fast track to innovation.

Applying pilots' lessons to doctors

Rather than resulting from a single fault, most major aircraft disasters are caused by a complex interplay between different human and technological factors. As a result (in addition to improving technology), decades of research have been conducted into human factors to identify the professional skills required of pilots. The commercial aviation industry has developed a mandatory Crew Resource Management (CRM) training programme to improve pilots' decision-making, teamwork and communication skills. The skills required of a pilot are very similar to those needed in an anaesthetist: complex decision making, situational awareness, communication and teamwork. The working environment is also similar: mostly long periods of monitoring but with the very real possibility of a crisis in which decisive action is needed. For this reason, the CRM has been used as the basis for anaesthetist training programmes in both the UK and the US. Stanford University has also used another aspect of aviation training, simulation, and applied it to the training of anaesthetists.

GSK and the Formula 1 pit stop mentality

GSK, the pharmaceutical company, applied the McLaren Formula 1 team's approach to quick changeovers to its toothpaste production line. Developing a pit stop mentality among its staff in relation to line changes halved the changeover time in its Maidenhead factory.

Making new connections between existing technologies

Putting different, existing technologies together to create something particularly inventive is very exciting.

The following case studies demonstrate how technologies may be combined to create something wonderful:

- **Trevor Baylis's clockwork radio:** Clockwork and radio had been around for a long time before Baylis thought to combine them. On its own neither technology is very novel, but combining them to create a radio that doesn't rely on external electrical sources or replaceable batteries was highly inventive.

- **Edwin Beard Budding's lawn mower:** The first lawn mower to be patented, in 1830, was inspired by watching the cutting cylinders used in textile mills to cut away irregular nap from woollen cloth to leave a smooth finish. Budding realised the technology could be reapplied to cutting grass, and developed the first lawn mower using a cutting cylinder with a series of gears, driven by a land roller. Another roller adjusted the height of the cut, and a tray captured the cuttings. Budding also invented the adjustable spanner or wrench – combining a spanner and a screw to create a tool that could be used with many different sizes of fastener.

- **Amazon's new way of buying books:** Amazon's great innovation was to combine three existing systems – bookshops, the Internet and the postal service – to create a cheap, speedy means of selling and delivering books.

What is interesting is that these great innovations didn't require any cutting-edge scientific research to prove them. A huge amount of work and technical expertise to develop them, sure; but the technologies behind the systems were already well proven. If you can systematically access the right, proven technologies to solve your problems, you can develop very innovative solutions with less risk.

Using the Database of Scientific Effects

When you want to know how to do something, TRIZ has the answer! You can look up the answers to all your 'How to?' questions in a database of scientific effects. Read on to find out more about this whizzy, TRIZzy tool.

Many of the TRIZ tools are useful for all kinds of problems. The TRIZ Effects Database is specifically designed for technical systems, and based on technical and scientific examples. Modelling your problem conceptually and looking for conceptual solutions is useful for all kinds of problems, and the simplest way to do so is to model it as an X-Factor (see the following sections to find out more).

Learning to look for what you want

The first step in finding what you want is learning to look for it intelligently. In fact, the step before that may be realising that perhaps you just need to locate it, rather than invent it.

When you have a problem, jumping straight to trying to dream up your own solutions is all too easy. However, someone else may have already tackled a similar problem and found a solution. The right way to start solving problems is to realise that what you need may already exist – either in whole or in part – rather than reinventing the wheel.

This is where the X-Factor comes in! The *X-Factor* is a simple TRIZ thinking tool for looking for solutions. When you want something, you imagine that some magical X-Factor can simply appear and provide it. The X-Factor enables you to look for the things you want either by making better use of resources (see Chapter 5) or as a standalone creativity tool (see Chapter 7).

What the X-Factor enables you to do is narrow your search for solutions that deliver the identified function you want – a delivered function is described as a simple Subject–Action–Object (see Chapter 12 for a full description of this process). You're looking for something (your subject) that's going to do something useful (the action) to something else (your object). In describing an X-Factor you're defining the delivered function you need – you then go on a hunt to look for all the best ways of achieving it.

You could describe what the examples in the earlier 'Making new connections between existing technologies' section were looking for as Subject–Action–Objects. In each of these cases, the subject is the X-Factor – the thing being searched for, as shown in Figure 6-2.

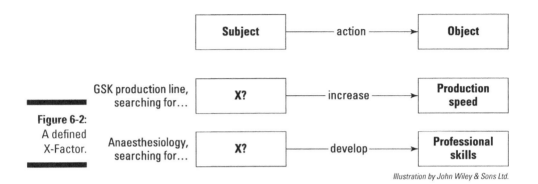

Figure 6-2: A defined X-Factor.

Illustration by John Wiley & Sons Ltd.

When you have a well-defined Subject–Action–Object, you can look up how to deliver your function in the Effects Database, consider resources (Chapter 5) or conduct an analogy search in the wider world.

Applying the Effects Database to solve problems

The *Effects Database* was developed over many years of research into patent records. Altshuller's idea was to take all the scientific effects and physical phenomena that the world had discovered so far (and captured in patents and scientific journals) and reorganise them in an inventive way, according to their application and usefulness – that is, what they do.

So when you have a problem you need to solve, you can go to the Effects Database and see what physical phenomena and scientific effects have been used in the past, and reapply them to your specific application.

The database was developed from the work and research of many scientists and engineers under Altshuller, and many versions are available. You can find condensed versions in various technical TRIZ books, in (very expensive) software and freely available on the Internet.

I particularly recommend two free online sources: Oxford Creativity's version (developed and maintained by Andrew Martin) at www.triz.co.uk and a Korean version at www.triz.co.kr.

When using the Effects Database, follow these steps:

1. **Describe your problem in normal language as a 'How?' question.**
   For example, 'How do I . . . purify water/carry water/ensure water is clean?'

2. **Define an X-Factor question to describe what you're hoping to do.**

   Your X-Factor will either define a function you're seeking or a parameter you need to change.

   Function examples: clean a liquid, constrain a gas, detect a solid, hold a liquid and so on.

   Parameter examples: measure purity, increase friction, change density, measure temperature, decrease volume and so on.

3. **Go to an Effects Database and look up the function(s) you want.**

4. **Translate conceptual solutions into real-world solutions.**

Clean car windscreen

A dirty windscreen as a result of mud, dust and bird droppings is a problem everyone encounters. You can think of some top-of-the-head solutions to this problem such as using your wipers or water and detergent to clean your windscreen, but what other solutions can you find, beyond your own knowledge, from the Effects Database?

Let's follow the steps above:

1. **Describe your problem: 'Clean my car windscreen'.**

2. **Define your X-Factor, as shown in Figure 6-3.**

Figure 6-3:
Defining an
X-Factor.

Illustration by John Wiley & Sons Ltd.

3. **Look up results in the Effects Database.** The results from the Oxford Creativity website (www.triz.co.uk) provide 90 solutions; these are: ablation, abrasion, acoustic cavitation, adhesive, adsorption, amphiphiles, brush, capillary action, capillary porous material, catalysis, cavitation, chemical transport reactions, combustion, composting, cryolysis, decomposition (biological), deflagration, desiccant material, desorption, electrical discharge machining, electrolysis, electron

beam, electron impact desorption, electropermanent magnet, enzyme, erosion, espresso crema effect, fan, fermentation, ferromagnetism, filter (physical), fluid spray, fractionation, friction, froth floatation, gettering, Halbach array, hydrodynamic cavitation, hydrogel, hydrogen peroxide, hydrophile, hydrophobe, ion beam, ion exchange, ion repulsion/attraction, jet, jet erosion, laser, laser ablation, light, liquid–liquid extraction, lotus leaf effect, magnetic field, magnetism, mechanical force, molecular sieve, nap, oxidation, ozone, phase change, photo-oxidation, plasma, purification, pyrolysis, radiation, radioactive decay, redox reactions, reduction, resonance, solvation, sonochemistry, sorption, sound, sponge, sputtering, sublimation, suction, supercritical fluid, supercritical fluid extraction, superhydrophilicity, surfactant, tribocorrosion, triboelectric effect, turbulence, ultrasonic vibration, vacuum, vacuum plasma spraying, vibration, wear, weathering.

4. **Translate conceptual solutions to real-world solutions.** In this case, many, many solutions exist. Some will be relevant, some less so, depending on your specific situation. Working through them to see if they suggest interesting and unlikely solutions is worthwhile.

The Effects Database will suggest a number of types of solution:

- ✔ **The ones you would've come up with anyway:** 'Fluid spray' and 'jet' both suggest conventional ways of cleaning, and 'amphiphiles' and 'surfactants' suggest using detergents.

- ✔ **Solutions you may know about but have forgotten:** 'Lotus leaf effect' is one of the self-cleaning glass technologies that you may or may not have come up with.

- ✔ **Solutions beyond your knowledge and experience:** These are the most valuable because they will take you beyond what you know to what the world knows: drawing upon a much larger pool of knowledge than any collection of experts you could assemble in one room.

The solutions you'd normally come up with depend on your expertise and field of knowledge. A mechanical engineer, for example, will tend to suggest mechanical solutions (fan, friction); a chemist will use chemical effects (solvents); a biologist will focus on biological solutions (enzymes); and a manager will delegate. The Effects Database will give you access to all these solutions and more: all the ways that have been discovered to clean a solid (so far!). Other solutions such as using vibration or ultrasonic vibration may not be obvious, but as they've made it into a toothbrush as a cleaning method, maybe they'll also be useful in this application. Lots of other solutions are suggested by the Effects Database, and if your business is making car windscreens, it's worth looking at them all!

The Effects Database gives you the answers that are most likely to help you. They direct you toward existing, proven solutions – both within and beyond your own knowledge. You must then use all your creativity and brainpower to turn them into practical solutions by stepping through your Prism of TRIZ, as shown in Figure 6-4.

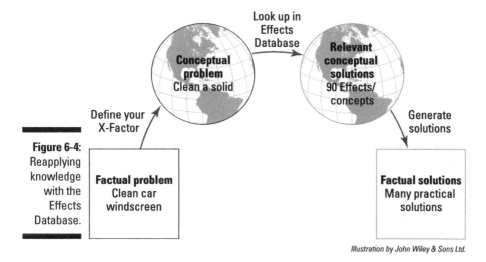

Figure 6-4: Reapplying knowledge with the Effects Database.

Illustration by John Wiley & Sons Ltd.

Strategies for analogous searching

If you have a technical issue to solve, you can look to the Effects Database for the function you need.

However, when you can't access the Effects Database because you're looking for a non-technical solution, you can search for analogous solutions using the strategies in the following sections. You look to other industries to see who's developed solutions that could become your X-Factor.

Life and death

The very best place to look for analogies is to consider for whom your particular problem is a matter of life and death. These people, organisations or industries will probably have the best solutions available.

Follow the money

Also consider those industries that have money to burn on research and development. They'll have investigated many potential areas and invested lots of money developing highly advanced technology.

Go cheap

Before it becomes a commodity, a product or system will have been developed and streamlined to be an elegant solution. Examples include pre-packaged food and drink, cleaning products, toiletries and toys.

Inventing with TRIZ

TRIZ was first developed to help anyone invent whatever they desired. Inventors are associated with myth and glamour, but what TRIZ really teaches you is that invention is just another form of problem solving.

We can all learn to invent. Starting with an existing invention and working out all the benefits it gives you is a useful approach. You then think about what other benefits it could be meeting that it currently isn't and create a wish list of these features. From these you identify your Ideal Outcome.

Matching needs and systems

All problem solving is about matching needs and systems.

A problem is just a gap between your needs – everything you want – and your current system.

Invention is problem solving because it means meeting previously unmet needs. You meet currently unmet needs with old technologies put into practice in new ways or in new combinations, or with new technologies you discover and develop.

As invention with TRIZ is simply a case of matching needs and systems, you need to understand both. You can then apply TRIZ tools and strategies to improve, as shown in Figure 6-5.

Analysing needs

Analysing your needs is as simple as defining your Ideal Outcome (see Chapter 9). You simply define all the benefits you want. Doing this in Time and Scale (see Chapter 8) is useful because your users' needs may change over time. Your Ideal Outcome will also help you scope your invention; that is, at what level you want to tackle the problem your invention will solve.

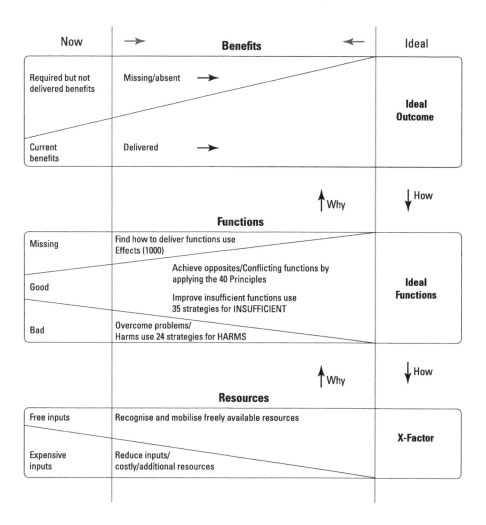

| Now | → | **Benefits** | ← | Ideal |

Required but not delivered benefits — Missing/absent →

Ideal Outcome

Current benefits — Delivered →

↑Why   ↓How

Functions

Missing — Find how to deliver functions use Effects (1000)

Achieve opposites/Conflicting functions by applying the 40 Principles

Good

Improve insufficient functions use 35 strategies for INSUFFICIENT

Ideal Functions

Bad — Overcome problems/ Harms use 24 strategies for HARMS

↑Why   ↓How

Resources

Free inputs — Recognise and mobilise freely available resources

X-Factor

Expensive inputs — Reduce inputs/ costly/additional resources

Figure 6-5: Understanding how needs, functions and systems connect.

System

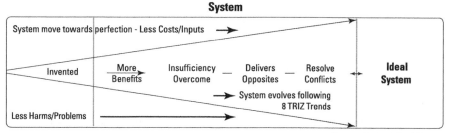

System move towards perfection - Less Costs/Inputs →

Invented — More Benefits — Insufficiency Overcome — Delivers Opposites — Resolve Conflicts — **Ideal System**

System evolves following 8 TRIZ Trends

Less Harms/Problems →

Illustration by John Wiley & Sons Ltd.

You may want to invent a method or device for locating a source of clean water, or a system for purifying water already collected, or a means of carrying clean water to where it's needed.

Analysing systems

Whatever you want to invent, someone has probably already invented something that doesn't do the job very well (or does so only partially).

One way to start inventing (as soon as you have your Ideal Outcome) is to take a real system and analyse it, understand all the ways it doesn't meet your needs, and then improve it. You can use any TRIZ problem-solving tools for this; what matters is to uncover all the problems within an existing system in relation to your new application and then to solve them one by one.

Say you decide to develop a cheap, portable system for purifying water for use following natural disasters. You could look at existing systems and uncover how they don't currently meet your needs (for example, UV light requires power, filtering doesn't catch all microbes, desalination is expensive and so on).

Understanding functions

The bridge between your *needs* and your *systems* is your *functions*, as shown in Chapter 5. Your systems have functions, which deliver benefits. Your benefits are delivered by functions, which are delivered by systems.

When you're inventing you can start with either needs or systems – making sure you've fully understood both – and the functions that could or should deliver them. Invention can start either with a system or a set of needs, but you must understand both the ideal state and the current state in order to problem solve and deliver a better invention – with clever functions. You can identify the functions you need and then see how you can achieve them using existing technologies (perhaps by consulting the Effects Database) and your resources.

For your water system, you may identify that the functions you need are:

- ✔ Kill microbes
- ✔ Capture sediment
- ✔ Hold water

Figure 6-6 shows how by identifying the functions you want (which deliver a benefit), you can then look for technologies or existing systems that can deliver these functions, one by one, and create a real invention.

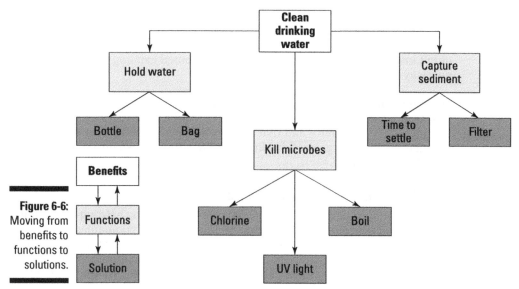

Illustration by John Wiley & Sons Ltd.

Figure 6-6:
Moving from benefits to functions to solutions.

Uncovering unmet needs

These needs may be obvious – we clearly need new solutions to meet some of the world's largest problems such as sources of water, food and energy – or they may be previously unarticulated or unrealised. Henry Ford famously said, 'If I had asked my customers what they wanted, they would have said a faster horse'. Check out the following examples!

Spanx: Understanding your users

Many famous examples of clever inventions result from people stumbling over an unmet need by chance, such as the invention of Spanx hosiery. Sara Blakely wanted tights without toes, but no one made them. She thus went into business developing her own (researching and finding companies to develop and deliver the product she wanted). Part of her success was testing her product on real people – and uncovering their unmet needs as a result. She discovered that, at the time, most products had the same size waistband (to cut costs), so instead she created a product offering a range of different waist measurements.

Windscreen wipers: Spotting an opportunity

In 1902, Mary Anderson saw snow for the first time on a visit to New York from Alabama. She noticed that trams had to stop every few minutes in order for the driver to wipe the snow off the windows. On her return home she invented and developed a squeegee on a spindle that could be attached to a handle inside a vehicle and moved to remove snow (rain, bird droppings)

from the window. She received the first patent for windscreen wipers in 1903 and within ten years this feature was available on thousands of cars; in fact, the same concept is still in use today, although greatly developed. Anderson didn't invent new technology: she simply combined existing technology in a new way to meet an unmet need.

Most inventions emerge from organisations rather than individuals. Researching, developing and testing an idea involves vast expense in terms of both money and time – and that's before you consider the cost of bringing it to market. When organisations invest in R&D, they're usually deliberately trying to meet a need.

The iPhone: Innovating smartphones

The smartphone is a famous recent example of combining existing technology to create a new invention and meet unmet needs. Many companies were developing smartphones in the early 2000s, combining the features of mobile phones and PDAs (personal digital assistants) that were popular at the time (digitally providing an address book, calendar, pad, calculator and various other features). One of the many reasons explaining Apple's great success with the iPhone was changing the user interface, creating a multi-touch screen that allows the user to type directly on it, thereby removing the need for a keyboard and stylus. This change in the user interface (and the very beautiful design for which Apple is famous) uncovered a need for smartphones to be easier to use. What's important to remember is that Apple didn't invent any of the fundamental technology to create the iPhone; rather, it made previous solutions work much better.

Kevlar: Using a new polymer in novel ways

Discoveries are usually made in organisations when people are already looking for something. Sometimes an unmet need is uncovered when a new technology is developed and people seek ways in which to apply it. Kevlar was invented by Stephanie Kwolek, a chemist who worked for DuPont. She was part of a team trying to develop strong, lightweight fibre to reinforce car tyres, but in the course of her research the material she discovered was much stronger than expected; it was five times the tensile strength of steel, but as light as fibreglass. Kevlar is now used in many applications, most notably in bullet-proof vests and helmets.

Applying existing technologies to create novel inventions

When you're trying to find new applications for technology, whether you've uncovered something new or want to find new uses for your existing technology, you start at the function level.

You then need to go up to benefits, to work out what needs could be met, and down to systems, to work out how you can turn them into reality. It's also worth looking at how functions can be put together to create a synergy, and potentially a new benefit, as shown in the real-life situations below.

Creating the ballpoint pen

László Biro, a journalist, noticed that the ink used in newspaper production dried quickly without smudging, and thought it would be useful to use for writing. He tried using it in a fountain pen, but the ink was too viscous to flow through the pen. Together with his chemist brother, Biro invented a pen containing a ball that moved freely in a socket and could pick up the ink and deposit it on paper. He took the already-developed quick-drying ink and reapplied it in a new system – inventing the ballpoint pen (and making a fortune in the process).

Reapplying NASA's research

Lots of clever inventions out there utilise NASA technology. For example, in the 1960s NASA developed memory foam to cushion aircraft seats. Following the release of this technology it's been used in mattresses, pillows, helmets for racing car drivers and American football players, footwear insoles, wheelchair seat cushions . . . the list goes on!

Part III
Thinking Like a Genius

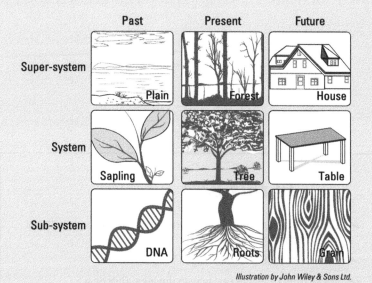

Illustration by John Wiley & Sons Ltd.

Check out www.dummies.com/extras/triz for a free article that explains how to help others boost their creativity.

Part III

Thinking Like a Genius

In this part . . .

✔ Get out of a mental rut and generate creative solutions with the TRIZ creativity tools.

✔ Learn to stretch your Thinking in Time and Scale.

✔ Travel from Utopia to reality with the Ideal Outcome.

✔ Develop a positive attitude to problem solving.

Chapter 7

Breaking Psychological Inertia with the TRIZ Creativity Tools

. . . .

In This Chapter

▶ Spotting psychological inertia and breaking out of it

▶ Enlisting the help of Smart Little People

▶ Using Size–Time–Cost thinking

. . . .

*I*f you want to learn how to think creatively on demand, it's important to understand what stimulates and what gets in the way of creativity.

We all fall into a mental rut from time to time, and in TRIZ we call this developing psychological inertia, which is one of the biggest blocks to creativity. All of the TRIZ tools help you challenge your psychological inertia, but before you can get out of your mental rut, you have to realise you're in it.

Psychological inertia is the TRIZ term for being stuck in a particular way of thinking. TRIZ gives you methods for stimulating your creativity as well as ways of examining your thinking, challenging your assumptions and thinking creatively. The TRIZ tools help you look at and understand situations in a new way and to find creative solutions to your problems.

In this chapter I also outline two powerful, standalone creativity tools in detail: Smart Little People and Size–Time–Cost. Both Smart Little People and Size–Time–Cost illustrate one of the fundamental TRIZ philosophies in action: start by thinking in extremes and then work out how to bring your ideas back to reality (more on this in Chapter 9). The logic behind this approach is that going to a far-out, wacky place restructures your view of the problem, allowing you to break out of psychological inertia and test constraints to find out whether they're real or not. You gain a new perspective on your issue and can think of new ideas without worrying about whether or not they're practical.

Ironically, thinking of wild ideas can be easier than coming up with practical suggestions, especially when you get into the habit of doing so. Mastering these tools will make it easier to free your thinking and think of all possible ideas. You can then turn these ideas into practical solutions.

Recognising Psychological Inertia

Psychological inertia is a term invented by engineers to describe a particular type of problematic thinking; it uses the language of physics to describe a psychological phenomenon. In normal language 'inertia' means doing nothing, but in physics this term refers to the resistance of an object to change – in motion, speed or direction.

An object can be moving, and will keep going in the same direction, at the same speed, unless another force acts upon it. This is a very insightful way of describing thinking, because getting stuck can mean your thinking is active but in a rut – you can't change direction.

You can fall into a mental rut because you're making assumptions, which limit your thinking. And the first rule of assumptions is that you don't know you're making them.

Your assumptions can define and narrow your thinking, so you no longer see what's going on around you in terms of the bigger picture. Often you don't even know you're in a mental rut to start with. So psychological inertia can mean you're stuck and can't move or it can describe active thinking that's being shaped by your attitudes, beliefs and biases.

When in doubt, assume you've got psychological inertia – and then challenge it!

The first step to getting out of your mental rut is realising you're in it. This is why so many of the TRIZ creativity tools challenge your view of problems, forcing you to take a step back and re-examine the situation. You do a reality check by continually asking yourself if you have psychological inertia; the answer is usually yes. For all of us!

Don't feel embarrassed about your psychological inertia and making of assumptions; these happen to the best of us. Without assumptions, comedians would have a hard time making a living because setting up and shattering them is the basis of many jokes. Here are some examples:

✔ 'They're always telling me to live my dreams. But I don't want to be naked in an exam I haven't revised for.' – Grace the Child

✔ 'I said to the gym instructor, "Can you teach me to do the splits?" He replied, "How flexible are you?" and I said, "I can't make Tuesdays."' – Tommy Cooper

✔ 'I've been in love with the same woman for forty-one years. If my wife finds out, she'll kill me.' – Henry Youngman

✔ 'I'm an excellent housekeeper. Every time I get divorced I keep the house.' – Zsa Zsa Gabor

✔ 'Outside of a dog, a book is man's best friend. Inside of a dog, it's too dark to read.' – Groucho Marx

✔ 'I want to die like my father, peacefully in his sleep, not screaming and terrified, like his passengers.' – Bob Monkhouse

These jokes are funny because the punchline's a surprise – and it's a surprise because the comedian telling you the joke has deliberately set up your assumption about the meaning of a word or phrase and then subverted it.

When people talk about 'insight' in problem solving and the 'aha!' moment when they generate a really clever solution, what they're usually describing is the realisation they're making assumptions and the point at which they think of a solution that breaks their psychological inertia. That moment feels fantastic and as enjoyable as laughing at the punchline to a really funny joke. You can experience that feeling and generate those kinds of solutions systematically by challenging your psychological inertia and applying the TRIZ creativity tools to generate clever solutions.

If you think you may be experiencing psychological inertia, you probably are. Keeping an open mind and reserving judgement on the nature of your problem and the nature of your solution is the best approach until you've worked through the TRIZ problem-solving process. Quite often, by doing so, you discover other ways of thinking about things and unexpected solutions that can surprise you.

I facilitated a problem-solving session for a local government department that was supposed to be about cutting costs in response to a reduction in funding. The team had already done a lot of work and found many ways in which to make savings, but nowhere near what was needed. The clever solution the team came up with was, rather than reducing costs any further, instead, to find alternative, external revenue streams to match and make up for the lost funding. This was a very innovative way of looking at the problem – challenging the psychological inertia that resulted in seeing cutting costs as the only way to manage on a reduced budget.

Every time someone says no, he's indicating he may be experiencing psychological inertia. 'You can't do that because . . . it's too expensive/it takes too long/it's too hard' are indicators that the person saying these things is making assumptions about the scope of possible solutions. He's also probably making assumptions about the specific way in which to put a particular solution into practice. You need to start with what you want and not make the assumption that it's going to be hard to get – this attitude could be psychological inertia at work. Just because you don't know how to do something, doesn't mean it's necessarily hard to do. It may be very easy for someone else – and very easy indeed to find out either way.

Many types of psychological inertia exist, including your

- View of the problem (how hard it is, how to tackle it, which parts are most urgent/important)

- Assumptions about the nature of the solution required (short-term, long-term, who the solution is for)

- Assumptions about constraints on a solution (time, money)

- Mental images or pictures of the situation

- Assumptions about what it's possible to change

- Tendency to get mired in the detail

- Previous success (which may not be possible to replicate in this situation)

- Previous failure (circumstances may have changed – things might work this time)

- Knowledge and experience (you view problems and solutions through the lens of your expertise, making assumptions about the nature of the problem and the ease of putting solutions into practice)

Appreciating the Benefits of Psychological Inertia

Although I'm being quite negative about psychological inertia, it happens for a reason. Psychological inertia is useful on a daily basis because its efficiency enables you to do tasks quickly.

Psychological inertia is necessary to get you out of the house every day. When you're getting dressed, for example, you don't need to think about it too hard, which is good because you can get dressed fast. When you put on shoes that require lacing, you don't think to yourself, 'How am I going to do this today?' Instead you just do it, quickly. In fact, thinking about automatic things consciously and trying to change them can feel uncomfortable – try putting your trousers on starting with the other leg to the one you usually begin with. Your habits become a kind of psychological inertia – and habits are hard to break.

Some tasks require explicit learning that then becomes second nature. When you first learn to drive, for example, you have lots to learn: starting with physical tasks such as how much pressure you need to apply to the pedals and how much to move the steering wheel to get the car to do what you want. You have to learn how to do these things first, before you can take on higher level problems such as merging with traffic and monitoring and predicting what other road users are likely to do. Initially, you have to think about these tasks consciously, but as you develop real skill, so you begin to drive automatically. In fact, if you later try to teach someone else how to drive, it can be remarkably difficult to remember what it is that your brain now does automatically.

When you're problem solving, your professional experience can act in the same way: because you've developed expertise, your brain can be efficient, taking shortcuts in thinking according to what you've seen in the past. This is the basis of your professional intuition; your experience has become second nature and unconscious. As a result, your brain has seen a particular type of problem many times before and works out the most likely, useful solutions to it without conscious thought. Unfortunately, your professional intuition is unlikely to lead you to creative solutions; it's hard-wired to repeat past success. When you're looking for new creative thinking, your expertise can thus get in the way.

Capturing the solutions that your experience suggests is nonetheless worthwhile. They'll be useful and may ultimately be the best solutions. But challenge yourself to think of new solutions too! The more experience and expertise you have, the stronger is your gut feeling towards trusted known solutions and the less flexible your thinking becomes, which reduces your creativity. The good news is that the more expertise you have, the more creative you can be when you explicitly apply creative thinking techniques and break out of your psychological inertia. It just requires the courage and discipline to try something new rather than applying the same type of thinking you've always used.

The point to bear in mind is that you're supposed to develop psychological inertia. It helps you think and behave efficiently, which frees up mental space and energy to think about higher level problems, and solve routine problems efficiently. However, psychological inertia can be a habit that's hard to break. It can load you down with misplaced assumptions or biases that direct your thinking in unhelpful directions and get in the way of creative thinking. Fortunately, TRIZ can help you overcome psychological inertia.

Developing competence and creativity

When you learn a new skill, such as golf, you're initially very bad at it, so bad that you don't even know it. You're unconsciously incompetent – you hit a golf ball and are thrilled when it goes anywhere at all and think you've done a great job. As you get better, ironically you feel that you're worse because you become consciously incompetent: you suddenly realise how far you have to go before you have a good swing and are able to manage the distance of the ball. Eventually, you move into conscious competence, when you've developed the skill but still have to think about it to get it right, mentally talking yourself through the steps required. Finally, you achieve unconscious competence when your mental self-talk diminishes and your swing is automatic. You can see these steps in everything you learn, at home and in your professional life.

When you want to think creatively, the last step – unconscious competence – is highly dangerous because you repeat behaviour that's worked in routine conditions. But to think creatively, you need to approach things in a different way, and you may be seeking creative solutions because routine conditions no longer apply. When you think creatively and use TRIZ, you move between unconscious and conscious competence: you allow your natural thinking to flow, but also check from time to time that you're going in the right direction and considering all potential solutions. Being directed by your unconscious competence may mean you're unknowingly blinkered by psychological inertia.

Beating Psychological Inertia

So you've realised that you're experiencing psychological inertia and you want to do something about it. But what? You explicitly look for and apply different ways of thinking: you are creative on purpose.

Why bad solutions are a good idea

One really important aspect of creative thinking is allowing yourself to think of all solutions without judgement. Most creativity approaches say much the same thing, including the rules of brainstorming (covered in Chapter 3).

One technique for making yourself view all ideas non-judgementally is to name them *bad solutions*, which simply acknowledges that they're imperfect and can be improved upon. Your, or someone else's, idea is a bad solution to the problem at hand; it's a top-of-the-head idea that can be improved and developed by you and others.

Bad solutions help you generate more ideas individually and as a team. Always capture all solutions as they occur to you, and put them in a 'Bad Solution Park'. Never squash any solutions, no matter how wacky they are (you can discard them later, if necessary).

Bad solutions are very important for creativity because they can break your psychological inertia. Bad solutions tell you something you want, and the reason they may be imperfect is simply that you don't know how to put them into practice (but someone else might). Bad solutions challenge what you think is possible, and you must always bear in mind that someone else may be able to find or already know a practical way to implement your terrible solution.

During a problem-solving session with a small team of software engineers who wrote the software for aircraft testing rigs, one of the problems they identified was the quality of the data they produced. Some wacky person suggested that the team buy a plane, which was so far beyond the realms of possibility that everyone laughed. But then someone realised they could actually rent a plane: a practical and cheap way to get what they wanted. This idea was ultimately one of five solutions taken forward from the day's problem solving.

If you have an idea, write it down. If you think it's completely impractical, you may just be experiencing psychological inertia!

Using the 13 TRIZ tools for creative thinking

Applying explicit TRIZ tools to shape your thinking on an issue will force you to look at it from a new angle, and also uncover the existence of psychological inertia and help overcome it. The trick is to allow this to happen – accept that your current view may be blinkered by assumptions and take a few minutes to explore it with a different way of thinking. Basically, you need to suspend disbelief or judgement for a time. All TRIZ tools challenge psychological inertia and stimulate your creativity in different ways so that you can apply it to your thinking throughout the problem-solving process.

That said, 13 of the TRIZ thinking tools and approaches can be applied as specific creative thinking tools:

✔ Thinking in Time and Scale (Chapter 8) helps you to explore at what level a problem needs to be tackled. Completing a 9 Box Solution Map (Chapters 8 and 9) helps you find inventive solutions to your problems.

✔ The Ideal Outcome (Chapter 9) helps you push the boundaries of what's possible and stretches your thinking to generate more ideal solutions.

✔ The Bad Solution Park (Chapter 11) helps you capture all solutions by challenging the nature of the solution required and stimulating new ideas by sharing solutions.

✔ Inventive Principle 13 (The Other Way Round; Chapter 3) shakes up your view of the problem to help you find creative solutions.

✔ The Prism of TRIZ (Chapter 6) helps you find analogous solutions: you strip out unnecessary detail, understand your essential problem and then find out if anyone has solved it before.

✔ The X-Factor (Chapter 6) helps you locate what you want using only what you have.

✔ Hybridising (Chapter 10) makes you combine solutions to get the best of each.

✔ Differentiating between benefits, functions and features (Chapter 5) helps you identify alternative ways to get what you want.

✔ Simple language (Chapter 6) helps you gain clarity of thought, which prevents psychological inertia regarding the scope of solutions.

✔ Idea–Concept thinking (Chapter 2) helps you generate many ideas from one solution.

✔ Life and death analogies (Chapter 6) make you consider others for whom your problem is critical and from there to identify relevant industries that may have tackled it in the past.

✔ Smart Little People (this chapter) shows you how to model problems and find new solutions.

✔ Size–Time–Cost (this chapter) helps you exaggerate your thinking to find new solutions.

All the TRIZ tools shown in this list can be used as simple creativity tools to help you restructure your view of the problem, stimulate your thinking to define your problem differently and find new answers.

Making creativity a new habit

One of the things that I find most inspiring about TRIZ is that it demonstrates that everyone can become more creative. You can make creative thinking a habit if you practise the right behaviours:

Be curious: You can't find creative solutions if you don't have any problems, so keep your eye out for them.

Try new things: Be open to new experiences, new ways of doing things and new thinking.

Challenge constraints: They may or may not be real, but you'll never know if you don't ask.

Make time for thinking: Taking time upfront to generate new ideas can seem like a luxury, but will ultimately save you time. When you compare how much time you spend putting something into practice and making it work with the time spent generating new ideas, the latter is usually squeezed. However, the stronger your solution, the easier it is to implement, and the fewer problems you'll run into later on.

Don't give up: Things will go wrong, but don't take this situation personally. Rather, it's an opportunity. You've found a problem! Apply TRIZ and find a solution.

Think big: Start with the biggest, best solution first and work down from there.

Be positive: Assuming that nothing's impossible generally means that's the case. Be positive about your own and other people's solutions and the possibility of change.

Understanding and Solving Problems Using Smart Little People

Smart Little People is a TRIZ thinking tool developed from the observation of clever and creative people at work.

When Altshuller and the early TRIZ community were first teaching TRIZ, they gave engineers and scientists problems to solve and asked them to talk aloud as they tackled them. They observed that sometimes people put themselves inside the problem; for example, if someone was looking at a leaking pipe he may say, 'If I was in the pipe, I'd cover the hole with a patch'. Altshuller noticed that when people put themselves inside the problem, they often came up with very inventive solutions. He thus developed a tool for creative thinking based on empathy, explicitly asking people to imagine themselves inside the problem area, wherever that may be – even if this was a physical impossibility in real life. What he then saw was that, while people came up with inventive solutions, if anyone else suggested things that would hurt or damage their imaginary selves inside the problem area (for example, using boiling water or acid), these solutions were rejected. The problem solvers experienced a form of psychological inertia: they wanted to protect their imaginary selves.

This problem was resolved by getting people to imagine a crowd of Smart Little People rather than just one person. In this situation they were able to imagine people being helpfully involved in solving the problem, but don't reject 'dangerous' solutions in case they get hurt. With an infinite number of people who can solve the problem, it doesn't matter if a few thousand get destroyed.

Modelling problems and solutions conceptually

Smart Little People provides you with a method to model your problem – and generate solutions – conceptually.

Smart Little People is a short-circuit around the Prism of TRIZ (see Chapter 6). You take a real-world, detailed problem and describe it as if the whole problem zone was made up of Smart Little People:

- ✔ **Smart:** Because they're clever. They can do – or be – anything. You can have helpful little people who solve problems and naughty little people who create them.

- ✔ **Little:** They can be as small as they need to be, right down to the cellular level (I've seen molecular bonds modelled as little people holding hands).

- ✔ **People:** They're not really people; they're pretend. But they have agency and can do stuff and work together as a team. Most people draw them as stick men, but don't let that hold you back – no reason exists for them to have two legs and two arms and one head.

Smart Little People is a very good tool for modelling any kind of real-world problem. I've encountered it used to model oil leaks in an engine (showing where and how the naughty little people escape and what can be done to prevent it); growing crystals for use in semi-conductors (showing how the molecular bonding is happening and where imperfections may be occurring); and medical devices (showing how a needle parts skin and exactly how the drug enters the patient's system) – to name but three.

When you use Smart Little People, you zoom in and enter the problem zone. As you model your problem, you identify exactly what's going on in that location. You become aware of the fine details so that you can focus on the place where your problem is occurring – but in a conceptual way. Your Smart Little People then help you find solutions.

Breaking out of practical thinking

Smart Little People helps bring very fresh thinking to a problem because you literally step out of the real world into an imaginary universe where you model a situation using teeny tiny people. It can generate quite wacky, silly thinking and, as a result, can be a lot of fun. For this reason, sometimes very serious people can at first be a little alarmed by the whole idea, but this aspect of the tool is actually quite important.

Studies demonstrate that creative thinking is often enhanced by a positive mood: if you're joking and laughing as you work on a problem, you're more likely to generate creative and inventive ideas than if you're all being terribly serious. A positive mood encourages flexible thinking, making novel and unusual connections between things and more fluid idea generation. However, in addition to the fun, Smart Little People is a structured and systematic way of helping you approach your problem from a new direction, and then to generate solutions. Because you're working in a pretend situation with imaginary people, it's easier to think of solutions to problems because your thinking doesn't have to be realistic and, as a result, is freer and more creative.

Let's step through an example, this time by redesigning a garlic press.

Traditional garlic presses are easy to use but fiddly and difficult to clean. You want to produce a garlic press that is easier to clean.

First, draw a picture of the system you want to improve or the problem area. Doing so helps you clarify what you want to look at and what you need to cover. Second, take your system and model how it functions using Smart Little People: model both what it does and all the problems that are occurring. Figure 7-1 shows one way in which your garlic press could be imagined as Smart Little People.

Third, when you've modelled all the problems as Smart Little People, think of ways in which useful little people could come along and help. What could they do to solve the problems? Model conceptual solutions as little people doing useful things: some examples are shown in Figure 7-2.

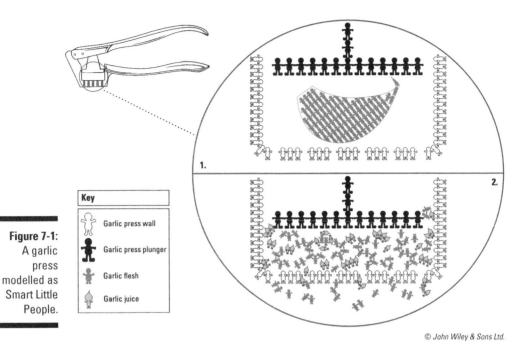

Figure 7-1:
A garlic press modelled as Smart Little People.

Key

- Garlic press wall
- Garlic press plunger
- Garlic flesh
- Garlic juice

During this step, don't worry too much about how your potential solutions become a reality. Instead, focus on how the little people could help. Figure 7-2 provides some examples:

- ✔ The garlic press people could push the garlic flesh and juice people away.
- ✔ Some other helpful little people could arrive with brushes and sweep away the garlic flesh and juice people.
- ✔ The garlic press people could chop up the garlic flesh people with swords.

The fourth step is to work out how these ideas could be turned into real-world solutions.

Bringing wacky ideas back to reality

When you've modelled your conceptual solutions with Smart Little People, you then have to work out how these could translate into real-world solutions. Your experience and expertise become useful at this point and you focus your search for practical solutions on a number of different specific functions that you've identified through the actions of your Smart Little People.

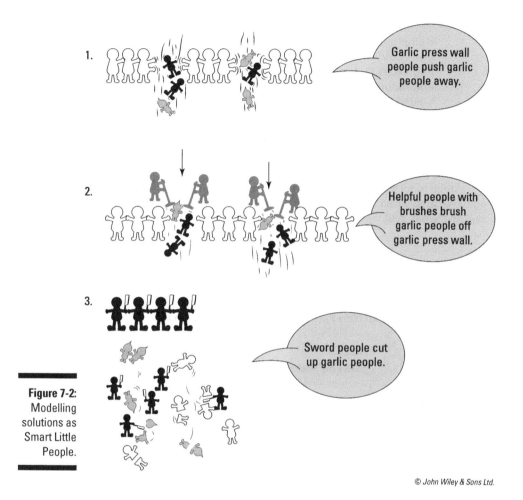

Figure 7-2:
Modelling
solutions as
Smart Little
People.

© John Wiley & Sons Ltd.

For the garlic press example shown previously in Figure 7-2, you take each of your conceptual solutions and work out how it could become a reality:

✔ You could imagine that the solution that involves pushing the garlic people could turn into some kind of coating on the press, such as Teflon, which makes the press very slippery, hard for the garlic to attach to and easy to clean.

✔ The new helpful people with brushes could suggest having another part of the press that slots into the holes and cleans them. It could be made of something that will clean effectively but not get dirty itself, such as prongs made from silicone.

✔ The swords idea suggests a method of chopping the garlic with something that's flat, sharp and easy to clean. Rather than crushing the garlic, you could create something that presses it against a set of sharp blades and slices it. Again, the surface needs to be easy to rinse clean, and, combining this idea with the one in the first bullet, perhaps made of or coated with something very slippery.

When you use Smart Little People, you move through the Prism of TRIZ in the manner shown in Figure 7-3. You start with a real-world problem, and translate it into a conceptual problem, described as little people. Then you imagine a conceptual answer, and have to translate it into something real. The last stage in moving around the Prism of TRIZ connects your conceptual, imaginative answer with your real-world experience and expertise to help you create eventual, real-world solutions.

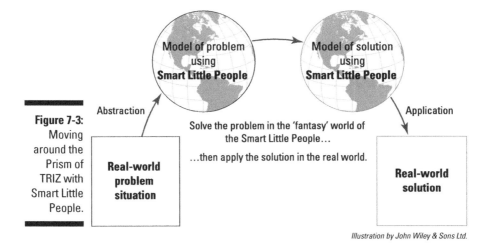

Figure 7-3: Moving around the Prism of TRIZ with Smart Little People.

Abstraction

Application

Model of problem using **Smart Little People**

Model of solution using **Smart Little People**

Solve the problem in the 'fantasy' world of the Smart Little People…

…then apply the solution in the real world.

Real-world problem situation

Real-world solution

Illustration by John Wiley & Sons Ltd.

Stretching Your Thinking with Size–Time–Cost

Generally in life you have a good idea of the practical limits of what you can do, what you can achieve and what you're looking for. These practical limits, however, may be hampering your thinking and causing you to make assumptions about what's really possible.

If you learn to stretch your thinking, you break psychological inertia and discover powerful new approaches.

Restructuring your view of what's possible

When problem solving or looking for new ideas, you generally have a good feel for the constraints of any solutions you come up with. When you buy a new car, you know roughly what sort of budget you can afford, even before you calculate the numbers properly. If a sliding scale exists ranging from very expensive cars such as Lamborghinis and Bentleys to more affordable options like Suzukis and Skodas, you probably know roughly where your budget fits on it. This is true of most situations in life when you're problem solving, even if you aren't consciously aware of it. That knowledge, of what you believe you can afford or is possible, can result in psychological inertia, whereby you place false constraints on the situation without realising you're doing so.

One of the trickiest aspects of psychological inertia is that often you don't know you're experiencing it.

Stretching your thinking and pushing constraints on elements such as size and budget is often worthwhile to see if spending just a little bit more can result in a cheaper option over time. For example, choosing a slightly more expensive but also more reliable car that requires less maintenance is probably cost-effective in the long run. You can apply that logic in both directions, so also consider spending less. Doing so may deliver as much as you want or need but you simply hadn't considered that option. When you work through this TRIZ process, you don't push your thinking a little, you think in extremes. You imagine what kind of car you'd have with both a zero *and* an infinite budget and a really enormous *and* tiny car, as shown in Figure 7-4.

Size

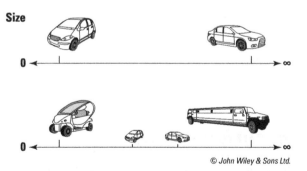

Figure 7-4: Stretch your thinking about the size of car you want.

© John Wiley & Sons Ltd.

Stretching your thinking in this way allows you to consider all types of solution beyond your initial assumptions and helps you come up with other ways in which to get the things you want. You may want a car that's suitable for a number of different purposes, such as taking long business trips, transporting large amounts of groceries and getting about town. Stretching your thinking so that you imagine an unlimited budget may suggest employing a chauffeur or buying a fleet of different cars, each perfect for its individual task. A zero budget may suggest not buying a car at all and using public transport. Bringing these ideas back to reality may suggest choosing a car most suitable for one task and finding alternatives for the others; for example, buying a small car for getting about town, hiring a big car for long journeys and getting all your groceries delivered.

The benefit of thinking in extremes is that you don't have to worry about practicality (at first). Often, when you're trying to think of solutions to real problems, you start to judge ideas almost immediately, testing them and picking them apart as they're generated. Doing so slows you down and can be discouraging. The luxury of playing with extreme, impractical ideas allows your brain to work most effectively because you're not constantly second-guessing yourself and instead your thinking can flow freely. After you've generated impractical ideas, you can then work out how to turn them into reality.

You stretch your thinking in this way using the *Size–Time–Cost* tool, as shown in Figure 7-5. These are the parameters that most commonly constrain your thinking, but you don't have to limit yourself to these. You may want to consider other parameters, such as number of staff if you're trying to solve a staffing issue. You can consider tackling the problem with an infinite number of people *and* no people at all.

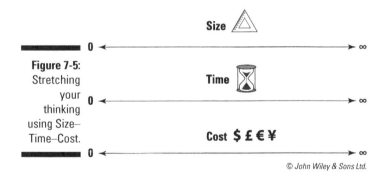

Figure 7-5: Stretching your thinking using Size–Time–Cost.

© John Wiley & Sons Ltd.

Finding inventive solutions

Consider trying to reduce energy use in an office building. Here's an example of using TRIZ creative thinking tools to generate innovative solutions.

Rising fuel prices and increasing environmental awareness mean that many organisations are now trying to reduce the amount of energy they use. Here we apply Size–Time–Cost to stretch our thinking and generate novel solutions to reduce the amount of energy being used to heat, cool and power offices. Table 7-1 demonstrates this process.

Table 7-1   Using Size–Time–Cost to Reduce Office Energy Use

Size	
Infinitely large	Build our own power station.
	Do something to the building as a whole, for example, improve insulation; change windows to reduce energy loss.
	Move office location to a more temperate region.
Infinitely small	Change office temperature according to the seasons and ask everyone to dress more appropriately as a result.
	Manage door and window openings more effectively.
	Don't heat/cool all parts of office.
Time	
We have forever	Design new building that requires less energy.
	Conduct full analysis of energy use to discover the source of most losses.
We have no time	Adjust thermostat up or down a couple of degrees (depending on whether it's cooling or heating).
	Move equipment and people around to manage heat better; heat or cool different parts of the office differently.
	Move to energy-efficient bulbs in all light fittings.
Cost	
Infinite budget	Buy or build a new building.
	Invest in research to identify equipment that requires less energy or is self-powered.
	Apply for grants to fund a partial move to renewable energy sources; for example, solar panels.
No budget	Find ways to use less energy; for example, make sure equipment is turned off when not needed (rather than placed on standby).
	Remove unnecessary or excessive equipment that consumes energy; for example, reduce number of fridges or printers; encourage people to take a paperless approach in their work.
	Make sure lights are turned off when rooms are not in use.

Chapter 8

Thinking in Time and Scale

*T*RIZ can show you how to think like a genius! One of the most exciting parts of learning TRIZ is that some of the tools provide you with a completely different approach to problems, shifting your thinking to help you look at situations with new eyes.

You learn to stretch your view of a situation over time and scale, delivering talented thinking – on demand. This chapter starts you off on that journey, using the idea of Thinking in Time and Scale.

Stretching Your Thinking in Time and Scale

Thinking in Time and Scale is a simple-to-learn-and-use TRIZ thinking tool. It emerged from the observation of clever and creative people (as described in the nearby sidebar, 'Learning to be creative again') as they solved problems. Thinking in Time and Scale is one way of modelling the natural thinking patterns of particularly creative people – and children – so that you can repeat them.

Everyone can think creatively, but many, if not most, people lose this ability as they grow up, as a result of being educated to think in a particular way. As a child you're taught to focus on the task at hand because doing so is very important when you're learning how the world works; keeping your thinking within sensible boundaries makes sense. However, this focus can form a kind of psychological inertia and place false constraints on your thinking and, therefore, your understanding of problems and situations.

Psychological inertia means being stuck in a mental rut (head to Chapter 7 for more details about this problem-solving affliction).

Thinking in Time and Scale helps you stretch your understanding of any situation you find yourself in; it challenges the scope of what you should be looking at and the kind of new solutions you ought to be seeking. Often, very innovative solutions can be found as soon as you get your head out of the detail, see the wood for the trees and think in time and scale.

What Altshuller and the early TRIZ community observed is that if you ask most people to think of a tree, they think of a tree. If you ask naturally creative people to think of a tree, they also consider the details: the roots, bark and leaves. They're also aware of the wider context within which the tree is situated, for example, in a forest. This perspective is rather like 'Google Earth-ing' a situation whereby you're able to zoom in and out of it by moving between levels of scale: zooming out from the system level to look at the bigger picture in the super-system and zooming in to look at the detail in the sub-system.

Learning to be creative again

Some of the earliest TRIZ research conducted by Altshuller and the rest of the TRIZ community involved observing people when problem solving with TRIZ. What they discovered was that about one in ten people are naturally very creative; that is, they're able to generate lots of potential solutions when faced with a problem. In an effort to understand what enabled this particularly creative thinking, these early researchers suggested that naturally creative people adopt two approaches that allow them to come up with lots of solutions:

✔ They initially think without constraints and allow all solutions to come through, no matter how wacky or impractical.

✔ They take a broader view of a problem, thinking not only of the present but also the past and future, and the impact of the larger environment and the fine details.

The thinking tool that helps you to think without constraints is the Ideal Outcome (see Chapter 9), but to take the broader view of a problem you need to learn to think in time and scale. Altshuller believed that we're all born creative but formal education suppresses that ability in most people. Those one in ten who remain naturally creative have managed to resist this process somehow, and now you can recapture that child-like creative thinking by applying the Ideal Outcome and Thinking in Time and Scale.

As soon as you've defined your system, everything above or outside it is the *super-system*. Sometimes this is also called the environment, but it refers to much more than the natural environment, and includes the structures and processes your system works within, things outside your system that it inter-acts with and intangible things like regulations and culture. The *sub-system* is all the individual parts that come together to make your system: all details of your system, technologies used, components. For example, if you took your organisation as the system, the super-system would include the markets you operate in, your competition, regulations; the sub-system would include your people, your products, your supply chain and so on.

As well as considering these levels of scale, naturally creative people are also able to think about how these levels change over time. They think in the present, refer to the past and look forward to the future. You can map these changing levels of time and scale in 9 Boxes, which is another name used for this way of thinking (for reasons made obvious in Figure 8-1!).

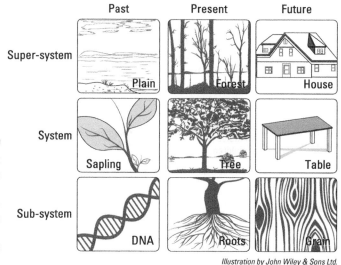

Illustration by John Wiley & Sons Ltd.

Figure 8-1: Thinking in Time and Scale – also known as the 9 Boxes.

Using 9 Boxes to map your situation helps you think creatively by helping you keep detail in its place (you need to understand the detail, but some-times it can overwhelm you) and finding new ways to view any situation. You can then actively stretch your search for solutions in both time and scale to find innovative places and methods to solve your problems.

Thinking in Time and Scale is a very straightforward but very powerful tool. Although simple, some people still find it confusing, however, because the

9 boxes can be used in many ways (some of those ways are described at the end of this chapter in 'Learning to Think in Time and Scale'). Another name for this tool is the 9 Windows, which I like because the tool resembles viewing a situation in different ways by applying a lens. Depending on the problem and the outcomes you're looking for, you use the tool in different ways (some of which are listed later in this chapter in the 'Learning to Think in Time and Scale' section).

Altshuller taught children TRIZ as well as adults because he believed that learning it as a child would help you retain your natural creative ability throughout your life. All children, Altshuller thought, demonstrated the same types of creative thinking and thought processes as the most creative adults, and he demonstrated this belief by setting children and engineers the same problem. The problem involved working out how to prevent a table in a nursery becoming damaged. Made of wood and painted red, the table was regularly subjected to the rough and tumble play of young children. Altshuller found that most people thought of pragmatic solutions near to the problem in time and scale: they suggested creating harder paint, which was more resistant to scratching, or repainting the tables. However, the particularly creative people and the children thought of solutions in many different times and at different levels of scale – which you can learn to do by thinking of solutions in 9 Boxes. Figure 8-2 shows examples of more inventive solutions that could be generated by Thinking in Time and Scale.

	Before	During	After Use
Children Kindergarten	Strict rules No tables No children	Nothing sharp used on tables	Punish children Create fashion for distressed tables
Table	Make of another, harder material (e.g. plastic)	Scratched red tables Use tablecloths/mats	Cover with tablecloths
Wood Surface of Wood Paint	Use harder paint Wood red all the way through	Self-healing paint	Repaint

Figure 8-2: Finding new solutions by Thinking in Time and Scale.

Remembering to stretch yourself

Despite the title, 9 Boxes doesn't constrain real-life problems into actual boxes! You are simply stretching your thinking. When you define the time steps, you can include in the left-hand column every detail from the moment before the problem you're solving occurred to the far and distant past right back to the creation of the universe. Looking forward, you can consider the moment after the problem occurred and beyond, to the distant future and even the apocalypse. In scale, you can zoom outwards as much as you want: you include everything outside your system, from the road to the physical location to the country. In the detail, you can zoom right into the molecular make-up of the issue at hand.

Usually your steps will be the past, present and future. The exact timings you pick will depend on what you're using your 9 Boxes for. As a general rule, however, the most important thing to do is decide on the boundaries between your time steps and levels of scale.

Of course, in reality you wouldn't go so far as back to the Big Bang or down to the levels of molecules – but remember to push yourself beyond the most immediate time and scale where your problem is occurring. You'll end up with a mixture of different time and scale steps in each box (for example, both near and distant future) – and that's okay. The point is to zoom in and out of the problem area to make you restructure your view of the problem and think in a different way.

Considering whether 9 boxes are really enough

But wait, I hear you cry, if I want to think about many different levels of scale and time, couldn't I just draw more boxes? The reality is that you can't cut up life into neat little segments. Every situation has a number of time steps and many levels of scale, but if you were to divide a situation into 81 boxes or more, you'd completely lose the plot. One of the benefits of 9 Boxes is that it provides clarity of thought, and the ability to see and track relationships over time and scale; using more than nine boxes means that clarity begins to get lost.

Psychologists suggest that people can hold in their heads and make sense of seven concepts, plus or minus two (more when you're well-rested and energetic; fewer if you're tired). Very, very occasionally it's okay to go up to 12 boxes, usually if you want to plan for the future and it makes sense to separate solutions for the near future and the distant future. However, if your situation demands more than that, for example, you want to chart a process that has nine time steps, I suggest you complete three different sets of 9 Boxes. Doing so means your thinking will remain clear and connections will be easier to spot and understand.

Completing the 9 Boxes: top tips

- Don't agonise – get going!

- If you're not sure if your scale or time levels are quite right, just get started because it will quickly become apparent whether they make sense. If you haven't got it quite right, just do it again; drawing four lines on a piece of paper is quick and easy!

- If you're uncertain about which box to put a detail or idea in, write it down anyway because you can always move it later; alternatively, write on sticky notes.

- Work systematically.

- Go through each box in turn and really think about what it means.

- Never leave a box blank.

- Start anywhere – just make sure you cover each box (but top left isn't a bad place to start).

- Think freely.

- As other ideas or thoughts for other boxes come to you, write them down and park them, but come back to the box you were working on. Sticky notes are useful for this: you can put them to the side and place them in the right box when you come to it.

Thinking big and thinking small

Thinking in Time and Scale teaches you how to think of the big picture *and* how to examine the detail. Most people have a preference – either finding it easier to think big or focus in on the detail of a situation. These kinds of thinking are equally useful, and Thinking in Time and Scale helps you learn to use both – and, even more importantly, see how they connect.

Thinking in Time and Scale is one of the TRIZ tools that helps you understand your situation more clearly by broadening the scope of your thinking. This can help you identify trends over time, connections between levels and new potential solutions: many situations are the result not of a single factor but rather a number of different factors that combine to create an outcome. When you're looking for ways to change a situation, you should therefore look in both the detail and the big picture for solutions.

Often you can find clever things in the super-system to deal with problems. British inventor Howard Stapleton, for example, invented a gadget called the Mosquito to deal with noisy teenagers hanging around outside shops. The device works by producing a very high-pitched noise that can only be heard by young people because the ear's ability to detect these very high frequencies declines as people age. This solution cleverly uses both the environment – as the noise is broadcast across the whole area outside shops – and the detail – by making use of how the ear changes over the lifespan: the noise is unbearable to teenagers but barely perceptible to most people over the age of 20.

Mapping out a situation in time and scale will help you identify new connections that can be made between the big picture and the detail, and to communicate these simply.

Looking at your situation with new eyes

Thinking in Time and Scale also helps you break out of psychological inertia and look at whatever situation you're in from a fresh perspective. Often you'll have a fixed view of what's causing a problem or at what level the right solution will be found. However, by mapping it out in 9 Boxes, you'll stretch your view of the problem and solutions, and be able to look at it with new eyes. Perhaps the perfect solution will be found by doing something in advance or letting a problem happen and dealing with it afterwards. Allowing yourself to consider the different possibilities means you free your brain and are able to think of things in a very different way.

One of the most important steps for ensuring you're working on the right problem is mapping out your situation in 9 Boxes. Doing so helps you restructure your view of the problem and ensure that you're taking a broad enough view.

If you regularly have a headache at the office, what's the right level at which to tackle this problem? Should you just take a painkiller? Mapping out the context of the situation is worthwhile: it could be in the detail (for example, you're not drinking enough water, you have eye strain) or in the big picture (for example, your posture, the computer screen settings). Mapping your situation in time and scale, as shown in Figure 8-3, will ensure that you're solving the right problem.

	Before	During	After
Super-system: **Wider environment**	Stressful journey? Uncomfortable journey?	Poor posture Badly set up work station At computer for 8 hours	Comfortable at home Less screen time More active Relaxed
Person	On way to work	At work with headache	At home, no headache
Sub-system: **Inside person**	Dehydrated? Wrong glasses?	Take painkillers Drink water	Drugs in system Rehydrated

Figure 8-3:
Context mapping a headache at work.

© *John Wiley & Sons Ltd.*

Coming up with great solutions to the wrong problem takes you further away from where you want to be – not closer!

Time is also a useful resource, and it's worth considering different time stages, what's happening then and how it can be useful. Effective supply chain management often utilises time in innovative ways, and many clever examples exist of using time required for one process to complete another.

One tasty example of using time as a resource is chocolate manufacturers using shipping time to make changes in the consistency of their product (for example, by adding an enzyme to it), so that when it's made and formed it's hard (which is easier to handle) but when it reaches the customer it's nice and soft.

The most common tools to use alongside the 9 Boxes are the context and solution maps. The context map is a useful early stage in problem solving to help you understand the scope of your problem (more detail on how useful it is in the problem-solving process can be found in Chapter 11). A solution map helps you stretch your thinking to come up with as many potential solutions as possible and can be used in conjunction with the solution tools; for example, the 40 Inventive Principles (Chapter 3), the Trends of Technical Evolution (Chapter 4), the Effects Database (Chapter 6) and the Standard Solutions (Chapter 13).

Understanding Problems in Time and Scale

Before you can solve a problem, you need to understand it. Mapping a problem in time and scale will make sure you've understood all its potential causes and ensure you're solving the right problem. People often have too narrow a view of a problem, based on their previous experience and limits in their knowledge or assumptions; by stretching your viewpoint you will ensure that you have pulled together all relevant problem information. Real innovation is often the result of an innovative view of the problem – and finding unexpected and clever places to solve it.

Recognising the importance of the big picture

Getting stuck on a detail when you're problem solving is all too easy. Sometimes when you're faced with a challenge you can over-focus on one small detail, which means you no longer see what's going on around you.

Thinking in Time and Scale ensures that you lift your head and look around your problem to make sure you're not becoming obsessed with one or two small details and losing sight of the big picture. It also ensures that you've scoped the problem correctly and are trying to tackle it at the right level.

Mapping causes of problems and hazards

When you're trying to prevent a very serious downside, Thinking in Time and Scale helps ensure you've mapped every potential cause of the problem or hazard. You can then systematically work through each of these problems and generate solutions to them. Thinking about problems can feel challenging and scary, but mapping them in 9 Boxes relieves the pressure by helping you scope out all potential causes so that you can then find clever solutions.

Many problems occur not when one thing goes wrong, but instead a number of unexpected circumstances come together to create a perfect storm. Each circumstance is a problem cause, and addressing just one of them is often sufficient to solve the whole problem. Many major disasters could be mapped in nine boxes and all their problem causes described; the more problem causes identified, the more opportunities discovered to prevent these problems re-occurring in the future.

The nine boxes in Figure 8-4 address the problem of barn owls being killed by road traffic. The problem occurs because these owls tend to hunt near roads, as they're a good source of prey; unfortunately, they're then struck by cars. This is a complex issue: the roadside is dangerous, but there are a number of reasons why barn owls loiter there. Fortunately, a number of clever solutions will be available to solve this problem and prevent the decline of the barn owl population.

Thinking strategically: Predicting future opportunities and threats

Thinking in Time and Scale enables effective strategic thinking. Whenever you're planning for the future, it's essential you consider how the world around you is changing, as well as the detail. This is important both from an organisational and a product or service perspective.

		Causes			Solutions		
		Before	During	After	Before	During	After
Environment		Barn owl habitats declining due to fewer barns; transport policy favours more major roads over other forms of transport	Road design doesn't consider barn owls; road verges encouraged as important wildlife resource	Little change in policy or road design in response to barn owl deaths; reduced owl population as roads act as barriers to young barn owl dispersal	Create more habitats at least 3km away from roads; no new unscreened roads in rural areas within 25km of resident barn owls	Design roads considering all wildlife; reduce traffic density and speed, perhaps only during autumn (when young owls disperse)	Restock barn owls in safe places; raise public awareness of risk to barn owls, especially in autumn; label 'barn owl blackspots' to encourage public to drive more slowly; collect injured barn owls
Road		Barn owl population searches for food within 3km of nesting site	More and faster cars and lorries; 3,000–5,000 owls hit by lorries per year	Lorries don't know barn owls have been hit; public not aware of what to do if they encounter an injured barn owl	Plant dense shrubs instead of grass on verges to discourage small mammals	No raised roads – sunken roads instead	Monitor and change roads where most casualties are spotted Train highways staff to identify bird species and record barn owl casualties
Wildlife		Young barn owls move to new areas in autumn; highest number of casualties in year	Barn owls fly low to hunt prey (like mice, voles) that can be found in verges near roads	Injured barn owls die on roadside; no official records of numbers of deaths	Collect and move young barn owls before dispersal to safer areas away from roads	Provide rough grass or outside obstacles/screens to encourage wildlife and provide food for owls	Create barn owl hospitals; teach local vets how to care for hurt barn owls; monitor barn owl deaths

© John Wiley & Sons Ltd.

Figure 8-4: Identifying the causes of barn owl deaths and possible solutions in nine boxes.

Your organisation in time

When you're thinking strategically from an organisational perspective, looking back to the past as well as forward from the present is helpful. This isn't navel-gazing – it helps you uncover patterns and trends for the future. Looking backwards at twice the timescale you intend looking forward is also worthwhile; for example, looking back ten years and forward five.

The exact timescales will depend on your organisation and those in your industry. An organisation in an industry with rapidly changing technology and ways of working, an app developer for example, will probably want to look at a shorter timescale (such as back five years, forward two) than one in an industry in which long-term projects are the norm, such as a defence company (for example, back 20 years, forward 10).

If your organisation has undergone any significant changes – for example mergers, redundancies or restructuring – it's worth using them to suggest the most relevant timescales for you. You need to think back to a time when you did things differently to how you do them now because this will help you capture any potential trends both within and outside your organisation.

Your organisation in scale

At the system level, you need to identify

- What you do or produce; that is, the reason for your existence
- Your core competencies
- Your structure
- Your turnover (and profit, if appropriate)
- Your location

Thinking at the super-system level, you should consider

- Who your customers are, their needs and locations
- Your markets
- Your regulatory environment

At the sub-system or detail level, you should consider

- The products or services you deliver; for example, the technology employed, range on offer, delivery service
- The type of people you employ, their skills, demographics and development; the rapidity of employee turnover
- Your suppliers

Developing your product/service strategy

Thinking in Time and Scale can also be useful for planning strategy for your main product or service. Rather than nine boxes, in this instance twelve boxes are better so you can consider both the near term and distant future. You still consider the past as well, because it will help you uncover trends and see new relationships emerge for the future.

One of the most valuable aspects of this exercise is seeing the relationship between the system levels, and considering the future not only of your product or service but also the environment it will be operating in and how the details and the super-system can interact.

So, when thinking in the super-system, consider the following:

- **Users:** Who will they be? How will their needs change? Where will/could they be geographically located?
- **Regulators:** How is the regulatory framework likely to change?
- **Partners:** How will they change? What do they need to do to implement your invention?

✔ **Supply chain:** What needs to change to successfully roll out your invention?

And in the sub-system, think about

✔ Trends in the technology you use (either within your product or to deliver your services) and potential future developments

✔ Competing technologies, products or services and what's holding them back at the moment (if constraints are removed/problems solved, could they leapfrog you?)

Completing this exercise will identify new opportunities for you in the future and also potential threats. Particularly when you're planning for the future, this exercise works very well with teams: it helps everyone understand your future goals and challenges, and can capture some clever and inventive thinking for the future.

Finding Novel Solutions in Time and Scale

Thinking in Time and Scale will stretch your thinking to help you find as many new ideas as possible. By broadening your view of where you can find ideas, your creativity is stimulated. You can also often find very clever and inventive places and times to solve problems – sometimes before they even occur.

Locating inventive times and places to solve a problem

Often the most efficient time to solve a problem is before it happens. If you can find a way to get the super-system to address the problem, you'll probably have found a very inventive and elegant solution.

Say you want to minimise the impact of flooding in your house. You can think of solutions at the super-system level, for example, not living on a flood plain; or at the system level, for example, changing the design of your house so that water naturally flows away; or at the detail level, such as keeping sandbags to hand. TRIZ, however, encourages you to look at what you can do not only to prepare in advance and deal with it when it happens, but also to minimise the downsides after the worst has happened. Considering the downsides suggests having tiled floors rather than carpet and locating your plug sockets halfway up a wall. One clever TRIZnik, having been flooded twice already, knew that one of the biggest downsides of a flood was not just getting rid of the water, but also the dirt the water brought in (especially if the sewers overflowed). When he heard that a third flood was imminent, he closed all the doors, put down sandbags – and turned on all the taps downstairs, flooding his house with clean water! He still had to deal with the flood – but the downside wasn't as bad.

Not everyone can be so courageous, but by mapping solutions in 9 Boxes you can often find very innovative solutions – and usually some you may otherwise have missed. While ideally you'd never reach the point where you have to engage these solutions, sometimes they're the most pragmatic options when problems are impossible to prevent. They can also provide a 'belt and braces' approach so you know that, even if the worst does happen, you can find ways in which to minimise the negative effects of the problem.

Covering the waterfront by finding all possible solutions

Sometimes you really need to demonstrate that you've thought of every possible solution to a problem. Thinking in Time and Scale really encourages you to think more broadly and deeply about a situation and prompts the generation of many more ideas than brainstorming alone. Most people, when faced with a problem, will be able to think of two or three solutions pretty easily. Drawing nine boxes, however, and using them to stimulate your thinking, encourages you to come up with at least nine – and hopefully more!

You can use the 9 Boxes with any of the solution tools to supercharge your TRIZ solution generation. Chapter 3, for example, considers the redesign of a cheese grater, and you could imagine applying the suggested Inventive Principles at different times: before, during and after use. You could also think about the levels of scale: in the big picture (the kitchen) or right down in the detail (the shape of the holes). Drawing nine boxes ensures that you generate as many solutions as possible and think of very novel ways in which to apply the Inventive Principles.

You can also arrange all the solutions generated by the end of a session in nine boxes and see if any gaps exist or some areas are light on solutions. Some people, for example, find it hard to generate solutions in the super-system, as they make assumptions about what's possible to implement. Generating solutions outside of what you think is possible is worthwhile because it will push those constraints, which may not be as impossible to change as you think. Considering what can be done after a problem has occurred is also useful because you may find clever ways not only to minimise problems but also to turn them into something positive.

Let's go through a simple example to practise generating solutions in time and scale. The common cold is a nasty viral illness experienced by most people. The cold is usually passed on in public places when infected people sneeze and tiny droplets of liquid containing the virus are inhaled by others. These droplets can also be transferred from skin or contaminated surfaces when someone touches her nose or mouth.

Plot out nine boxes and think of the solutions you already know about; then stretch yourself to think of as many more as possible. Don't feel constrained by reality: think of ideal as well as more pragmatic solutions.

Define your time steps and level of scale first.

You can prevent a cold in many ways. Figure 8-5 shows a number of potential solutions that you may have thought of, including the practical (stay warm), those outside of most people's control (increase public funding), the impossible (develop vaccine) and the downright wacky (financially reward sick people for staying at home). Capture them all – the more solutions you generate the better. These solutions may contain something clever that can be partially implemented (see Chapters 10 and 11 for more on why bad solutions are a good idea).

		Before	Exposed to bugs	Develop illness
	Super-system: Wider environment	Live somewhere warm Eradicate common cold Increase funding into research Educate public about good practices for avoiding spread of cold	Ensure good air circulation on public transport More public transport to prevent over-crowding Disinfect door handles/bannisters/taps etc. in public buildings/workplaces Don't shake hands with people in winter	Isolate ill people Discourage/prevent people from going to work and spreading germs Financially reward people with colds to stay at home Take swabs from sick people to help develop vaccine
Figure 8-5: Preventing the common cold in time and scale.	Person	Maintain physical health and fitness Eat a healthy diet Don't get overtired and stressed Avoid enclosed spaces with many other people in cold seasons, for example, public transport	Wash hands and use antiseptic hand gel regularly Don't touch eyes, nose Wear gloves on public transport Keep nose warm (with scarf) Keep body warm	Sneeze into tissue, then discard; or crook of arm, not into hand (or air or on someone else!) Maintain good hygiene practices Rest Monitor sick people for complications
	Sub-system: Inside person	Develop vaccine Strengthen immune system	Prevent bugs multiplying inside body Strong immune system fights bugs Stay hydrated Inside of nose impervious to bugs	Take zinc: reduces duration and intensity of illness Stay hydrated Take medication (like paracetamol) to reduce symptoms Ease symptoms with hot drinks, decongestants

© John Wiley & Sons Ltd.

Thinking in Time and Scale gives you a new perspective. You challenge your view of the problem and encourage talented thinking. Not only does thinking in this way help you find new solutions but it can also provide great clarity of thought by showing the relationship between the detail and the big picture – and how it changes over time.

Practise generating solutions in 9 Boxes by imagining real downsides you'd like to prevent in your life (for example, a fire in your home or a car crash).

Learning to Think in Time and Scale

The best way to learn how to do something is to put it into practice.

This section offers up my best tips to get you started on Thinking in Time and Scale:

- ✔ To understand your problem better, complete a context map. This is useful at the beginning of problem solving because you question how you arrived in your current situation and where you could end up. It's also useful for defining strategy and uncovering potential opportunities or threats for the future.

- ✔ To find new solutions, complete a solution map to generate lots of creative ideas.

- ✔ To uncover all potential causes of a problem, complete a problem causes and hazards map.

- ✔ To understand your process better, complete a process map to work out both the details of your process and how it interacts with the environment.

- ✔ To locate all available resources, use a resources map (see Chapter 5 for more detail on resources).

- ✔ Capture how an ideal system would operate at different times and different levels of scale to define an Ideal system (see Chapter 9 for more on the Ideal).

- ✔ Sometimes requirements change or are in conflict across levels of scale or at different times. Capturing a requirements map in nine boxes helps you pull out potential contradictions and understand everything you really want.

Chapter 9

Living in Utopia (then Coming Back to Reality)

he *Ideal Outcome* is the TRIZ tool for defining requirements quickly and clearly at the beginning of any project or task – no matter how big or small. The Ideal Outcome is a utopian mindset that changes how you approach problems and solutions, and is one of the most powerful tools for encouraging clear and creative thinking.

One of the really important aspects of the Ideal Outcome is that it enables you to think of *all* the things you want. It's a bit like thinking about what you'd do if you won the lottery: what you'd do if money were no object . . . but not just money – ignoring any constraints of practicality. Setting your Ideal Outcome is a very powerful exercise for clear thinking. Clearly defining what you really want drives innovation and often enables you to find something that more perfectly matches your needs than you'd imagined.

Defining the Ideal Outcome

Your Ideal Outcome is an essential first step in understanding what you want to achieve, defining not only your requirements but also the direction and scope of your problem solving.

Locating your North Star: Setting your problem-solving direction

The Ideal Outcome is how you define your needs and capture all requirements with TRIZ. Your Ideal Outcome is a wish list – everything you'd have in
an ideal world without specifying (yet) how you'd achieve the outcomes. This
ideal thinking has a practical purpose: while you never expect to achieve
everything on your wish list, identifying all the things you want ensures that
you work towards the things that, like a Spice Girl, you really, really want,
rather than just the things that you think are possible. The Ideal Outcome is
independent of any system – and many ways of delivering it will exist.

Too often when people are asked what they want, they offer you a solution –
one way that they can imagine of getting the thing that they really want.
Very often in practical problem solving, you uncover the real problem when
you start defining the Ideal Outcome; you discover that the problems you
need to solve are actually problems with someone's imperfect solution that
was put into place to solve the fundamental, underlying problems.

Defining what's 'Ideal' frees your thinking from practicality and sets the
direction of your problem solving: by allowing yourself to think of where you
really want to go, you break out of thinking in constraints and within the systems you're currently working with, and allow yourself to lift your head and
see the big picture. So in a way, defining your Ideal Outcome is like the 'North
Star' for your problem-solving journey. You'll never actually get there – just
as you won't actually arrive at the North Star – but it tells you the direction
you ought to be going in.

An Ideal Outcome has no solutions – it's a list of benefits. A benefit is an
outcome: it doesn't tell you *how* – it just tells you *what* you want. There'll be
many, many ways of achieving any benefit.

For example, if I were to say I wanted pockets on every item of clothing I own,
that's a premature *solution*. The *benefit* behind having 'pockets' is *fast and easy
access to belongings* such as keys, wallet and glasses. Other ways of getting
those benefits exist – I could carry a bag. Or put my things in someone else's
pockets. Or not need to carry any stuff. Focus on benefits, not solutions!

Outlining benefits as a team

Talking about benefits and not solutions enables open communication
between teams – and different specialities. People often get stuck in the detail
of how things are delivered, and can make assumptions about how much is
possible based on their own (possibly limited) knowledge.

The importance of defining everything you want

A manager and two of his employees are out to lunch and stop off at an antique shop. They spot a lamp and start dusting it off when, to their surprise, a genie appears in a puff of smoke. He offers them three wishes – one each. One of the employees wishes to live on a mountain, so he can go skiing whenever he wants. The other wishes to live on a tropical island, far away from civilisation. And the manager? 'I want those two back at the end of their lunch break,' he says.

Spending time understanding what you really want is important – you need to clearly explain what you want and capture *all* of the things you're looking for.

Try to set the Ideal Outcome as a team: sometimes you need team consensus on the direction in which you ought to be heading, and doing this together allows everyone's voice to be heard. It also ensures at that first early and crucial step that you've all agreed the scope for the problem-solving session. The Ideal Outcome allows you to do this quickly – because it's ideal and you don't expect to get there, it allows everyone's wishes to be heard and captured (although you know you may have to make decisions to choose between them later on) – and ensures that all the requirements have been identified.

I often give people actual little magic wands when we're defining the Ideal Outcome because it helps emphasise the fact that this isn't your usual type of thinking. It's also good if people are having fun and laughing, as a positive mood helps you think more broadly and without constraints, which is an important part of the process. When you hold the magic wand, you allow yourself to step out of the real world and imagine all the things you want. When you put the wand down, you can then start thinking in terms of constraints and being practical. You can harness that free and open thinking early on, knowing that it's only one stage of the process; the pragmatic thinking comes later on. Holding a wand encourages free thinking, as the pragmatists won't feel they need to put the brakes on the requirements too soon.

Here's how to define your Ideal Outcome as a team (see the next section for an example):

1. **Ask everyone to think about what he wants in an ideal world, concerning the project or task.** Write one wish per sticky note (big pens are helpful, as they encourage people to be concise). Encourage free thinking, without categorisation or ranking, just all the benefits they'd have in an ideal world, as many as they like.

2. **Get everyone to stick their notes on the wall.**

3. **Ask the team to group similar notes together – without talking.** Everyone should be participating, and doing it silently stops any one individual running the process.

4. **Come up with names for the groups of notes: these are your benefits; say, 'portable' or 'easy to use'.**

5. **Organise the notes into sub-groups of must-haves and nice-to-haves.**

6. **Identify the Prime Benefit (the one main thing you want).** This may not have been captured.

When you're first trying to define your Ideal Outcome, you can start with initial thoughts and then turn them into more ideal thinking by asking 'why?'. Challenge every statement on your wish list and ask yourself 'why do I want that? What's good about it?' This challenges you to uncover all the benefits you really want – and prevents you from jumping to premature solutions.

Later on, you come back to reality by thinking about generating solutions (in the later section 'Taking a Step Towards Reality with Ideal Systems').

If you are using TRIZ in conjunction with another toolkit and want to use another tool for defining requirements, I urge you to do an Ideal Outcome first as well, just to stretch your thinking. It doesn't take long, only 20–30 minutes, and you can then feed in the relevant information to your toolkit. Many other approaches muddle benefits and functions and doing your Ideal Outcome ensures really clear thinking.

Striving for perfection with the Ideal Outcome, Prime Benefit and Ultimate Goal

Defining the Ideal Outcome not only helps you define the perfect world but also, somewhat ironically, the constraints. The Ideal Outcome should be matched to the task in hand. Defining an Ideal Outcome for a new product is very different to defining one for a fix/mend problem. Spend time at the beginning of whatever problem you're tackling, scoping it and exploring whether you're looking for a quick fix or a complete redesign (any good TRIZ session should generate both kinds of solutions) so you can consider where you want to focus your attention. Your Ideal Outcome therefore not only breaks you out of thinking in constraints but also helps you identify what those constraints ought to be.

Here's an example. One invention I'd really like to create is an amazing water bottle. Having water when you're out and about is useful, but it seems wasteful to keep buying bottles of water. I've tried a number of reusable water bottles but they've all been imperfect (so far).

What would all my requirements for a water bottle be?

Wish list/benefits:

- ✔ Can carry an infinite amount of water
- ✔ Always clean (inside and out)
- ✔ Can carry any liquid (squash, coffee, fizzy water)
- ✔ Easily portable (zero weight, zero volume)
- ✔ Keeps water cold or coffee hot

Of course, that list is impossible. The Ideal Outcome stretches your thinking – the only numbers that you need in an Ideal Outcome are zero, 100 per cent or infinity.

Sometimes it's worth writing down what you want and then stretching it to be ideal. For example, first of all, I thought an ideal would be to 'carry a lot of water' but 'a lot' is not ideal. An ideal water bottle would always have as much water as I want – an infinite amount. That's impossible, but it breaks me away from the psychological inertia of deciding how much a 'reasonable' amount of water is (500 ml–1 litre, for example). Solutions may be out there that could carry a lot more water than I think – but they might not occur to me if I'm already considering constraints in size and volume.

After you've identified the benefits you want, you identify your *Prime Benefit* – the one thing your system exists to deliver. Sometimes this will have been captured in the benefits you've listed, sometimes not: that's okay – just make sure you capture it.

> **Prime Benefit:** Portable container for carrying water

Then you define the Ultimate Goal – why are you doing this?

> **Ultimate Goal:** Access to clean drinking water on demand

Defining your Ultimate Goal is the point at which you have a discussion over what single goal you're trying to achieve. If, instead of *a portable container for storing water*, you want *access to drinking water on demand*, you could consider providing water fountains, or create a subscription model where participating restaurants and cafes give water on demand, or carry a LifeStraw® (see the nearby sidebar, 'LifeStraw: Clever resource use').

As you discuss your Ideal Outcome, you start to define the scope of solutions you want to work on. If you decide you actually want to create access to free, safe drinking water for everyone, you'll be developing very different solutions to the problem of creating a portable container for storing water. Pinpointing your Ideal Outcome means you sometimes discover that the problem you're working on is actually an imperfect solution to a deeper, underlying problem.

LifeStraw: Clever resource use

The LifeStraw is a very clever TRIZzy invention. One of the biggest problems the developing world faces is access to clean water: diarrhoea resulting from drinking unclean water kills more than 1.5 million people per year. Even when clean water is available, it requires transporting and is very heavy. The makers of LifeStraw realised that what people really want is access to safe, clean drinking water on demand. Storing and transporting clean water from a safe water source is only one way of getting that. They created a portable straw that filters out parasites and kills bacteria, creating clean water from dirty water (which is often readily available). So the need to store and transport water completely disappears: the LifeStraw uses an existing resource (dirty water) to give people what they want. LifeStraws are available for hikers and campers, and have been distributed after major disasters.

Setting your constraints

Your Ideal Outcome helps you consider all the things you want without worrying about how you get them – you see new possibilities that might have eluded you if you'd started with pragmatic thinking. However, now you take steps towards reality by defining your constraints and what solutions you want to develop. The Ideal Outcome points you to where you want to go.

This is the moment to set your constraints, and the most important thing to consider is what sort of solutions you're able to implement. Generating clever solutions to give the whole world access to safe water is pointless if you don't have the ability to put them into practice! Putting a team through an exercise that ends up being theoretical will tire and annoy them, and make them reluctant to participate in any future sessions.

Constraints to consider are

- The ability to implement solutions
- Timescale for delivery of solutions
- Budget available
- Required life of system
- Regulations
- Physical limitations, such as size and weight

Sometimes needs face contradictions or conflicts – either in the things that you want or between different stakeholders. It can be tempting to think that these conflicts are impossible to resolve – but don't give up too soon! Start by believing that you'll be able to get everything that you want – and if you face a contradiction, turn to Chapter 3 to work out how to solve it.

TRIZ gets the party started!

When I moved into a small flat, I did my Ideal Outcome for everything I bought (TRIZ-nerd alert!). One challenge was wine glasses: I wanted a lot of wine glasses because I like having parties, but I had limited storage space. I had a contradiction. However, rather than giving up, I looked for a solution and discovered stackable wine glasses. Defining your Ideal Outcome can help you find very clever and elegant solutions to your problems, and also stops you from compromising too soon.

Considering stakeholders: Don't be afraid to ask your customers

I often recommend that when people are defining the requirements for a new product, they ask their customers what their Ideal Outcome would be. People usually recoil in horror at the suggestion, stating 'I can't give them a magic wand! They'll ask me for a whole load of things I can't deliver!' Well, maybe – but when you find out what your customers really want, independent of what you currently do – you may well discover better ways to deliver it and new places to innovate. In problem-solving scenarios, participants often observe that they fail to understand customers' needs because they have difficulty articulating them. Too often customers ask for a specific solution, rather than expressing a need. They assume that only one way to get what they want exists, whereas their suppliers may know other ways. Understanding these needs may thus give you the opportunity to innovate. This is particularly important when, as a customer, you are purchasing systems that you don't understand or for which developments are fast, such as software. When choosing, developing or changing IT systems, it's very important to have a clear discussion of the benefits required and constraints faced. Often when you change something you focus on one new benefit but may inadvertently lose benefits provided by the old system (and not realise it until it's too late).

Benefits may also exist that your customer wants but you didn't realise were possible, which offers a fantastic opportunity. Sometimes your customers want something but, because they lack your product knowledge, never ask for it because they assume it will be too hard to deliver. Actually, you may find doing so really easy. Creating an Ideal Outcome frees everyone's thinking and allows your customers to reach for the moon.

You may also be delivering things to your customers about which they feel indifferent. Over time, providing these things has simply become the norm, which is particularly true of legacy projects or long-running lines of products. For clear thinking, however, it's essential to check what's needed for every specific project to ensure that you don't unnecessarily over-deliver.

Agreeing on Ideal Outcomes with your customers isn't scary because they're ideal, not reality. You can follow them up with sensible discussions about how much you can deliver and at what cost. See the 'Making Sensible Decisions by Considering All Benefits, Costs and Harms' section, later in this chapter, for more on cost.

Taking a Step Towards Reality with Ideal Systems

The Ideal Outcome is very much about freeing your thinking and uncovering all requirements. The next step is to bring that powerful thinking a step closer to reality. The following sections explain the logical steps you can follow to focus your thinking into practical and real systems, and consider not only the good things you want, but also the downsides you are willing to tolerate.

Thinking of what you want and then getting it

One of the most powerful aspects of thinking of the Ideal Outcome is that it helps you uncover what you really want. By focusing on the Ideal Outcome – of benefits without solutions – you often discover that what you want is quite different from what you thought you wanted. Sometimes, simply defining the Ideal Outcome can result in you suddenly seeing a solution that's eluded you for months, because you hadn't recognised the underlying benefits you wanted. Sometimes imagining the Ideal Outcome also enables you to see how to get it. Maybe someone else has already created what you want, or clever alternative ways to get it exist.

When you've established your Ideal Outcome, you then have to bring it back to reality and start seeing how to achieve it. One way of doing so is to take existing systems or any of the bad solutions that you've generated and work out what they lack. Another is to identify the benefit you most want and work out what gets worse as you gain more of that benefit. To return to the water bottle example (in the earlier 'Striving for perfection with the Ideal Outcome, Prime Benefit and Ultimate Goal' section), the thing I want most is large capacity. However, as the bottle gains greater capacity, so it gets bigger and heavier, and I also want it to be easy to carry around. Herein lies a contradiction (see Chapter 3 for more on solving contradictions).

When you identify a contradiction, you have a route into generating solutions. Considering Ideal Systems is another way to prompt the creation of solutions.

Using 9 Boxes to define your Ideal System

Defining your Ideal System is a good way to keep thinking broadly about what you want while also taking a step towards a real solution. Thinking in Time and Scale with 9 Boxes is a useful way to stretch your thinking and encourage innovative solutions; see the example in Table 9-1. I've used this approach with clients on every kind of system, from oil rigs to medical devices.

Table 9-1	Ideal Water Bottle in 9 Boxes		
	Before Use	*During Use*	*After Use*
Super-system	Clean water available No storage space necessary	Easy to carry/ transport (no weight)	No environmental impact (it doesn't end up in landfill or pollute)
System: Water bottle	Robust Easy to fill Easy to store and find	Holds a lot of water Easy to drink from	Takes up zero space to carry home
Sub-system	Clean water Interior of bottle clean	Water safe to drink	Water is inside person Interior of bottle clean

Consider how an ideal water bottle would function in terms of scale and over time (for more information on the concepts of time and scale, check out Chapter 8). How would this bottle work as a system?

The next stage is to use the Ideal System to start creating some practical solutions. You challenge yourself to come up with ideas so that you can make your Ideal Outcome a reality. In the case of the water bottle, the Ideal System suggests that what's wanted is a vessel that's small for the sake of storage, then large enough to contain a lot of water when in use, and then small again to transport home. Clearly, a Physical Contradiction exists, which you can resolve by separating what you want at different times.

Using divergent thinking with Ideal Functions

When moving from the Ideal Outcome to the real world, thinking in functions rather than moving straight to systems is the key stepping stone. Functions are provided by systems (our solutions) and are the 'how' – how you achieve the benefits you want. You need to list your benefits and then think of functions that can deliver them (more than one function may deliver each benefit). When you've identified the functions you want, you have a starting place for generating solutions/systems. You should be able to think of many potential solutions when you consider each function. Table 9-2 provides potential solutions.

Table 9-2	Identifying Ideal Functions	
Benefit	*Function*	*Solutions*
Can carry an infinite amount of water	Holds water	Bottle
		Bag
		Shell/bowl
		Sponge
Easily portable (in use)	Carries water	Bag in hat/belt/backpack/bra/jacket
		Strap for bottle
Easily portable (after use)	Bottle shrinks	Bladder/bag
		Collapsible bottle
		Telescoping bottle
		Folding bottle
		'Water balloon' bottle stretches under weight of water, returns to small shape when empty
Easily portable (after use)	Bottle disappears	Edible, biodegradable, heat degradable, waxed cardboard
Always clean	Easily washable	Material dishwasher-proof
		Unfolds to lie flat/dismantles for easy cleaning
		Opening enlarges to wash inside
		Microwavable to sterilise
	Repels dirt	Made of silver
		Silver coating
		Antibacterial coating

Table 9-2 considers only a few functions. Other elements to consider are how to make the vessel robust to avoid damage and how to close and seal it.

Another approach to generating practical solutions is to create an Ideal User Manual. It's an excellent way to think about your system from a user's perspective, which possibly makes you derive more elegant solutions and remove steps that the user may find onerous. It should also start to trigger practical solutions.

Making Sensible Decisions by Considering All Benefits, Costs and Harms

Your Ideal Outcome can form the basis of very sensible and pragmatic decisions by considering your Ideality – the ratio between the benefits your system delivers, the costs you have to put in and the harms that result.

Ideality is the TRIZ way of thinking about value. But rather than considering costs and benefits alone, you also think about harms. The Ideality is the measure of how good you judge your system to be, in terms of a ratio of its benefits over its costs and harms, as shown in Figure 9-1.

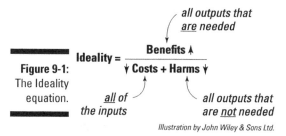

Figure 9-1:
The Ideality
equation.

$$\text{Ideality} = \frac{\text{Benefits}}{\text{Costs} + \text{Harms}}$$

all outputs that *are* needed

all of the inputs

all outputs that are *not* needed

Illustration by John Wiley & Sons Ltd.

A cost is anything you put into your system. Cost in TRIZ refers not only to time but also money, effort and the world's resources – any input, in fact. Your inputs give you two outputs: benefits and harms. No neutral outputs exist in TRIZ terms: they're either good or bad.

A *benefit* is an output of your system that you want; a *harm* is an output you don't want.

TRIZ is brutal about harms. It finds them everywhere. A harm can be heat from a lightbulb, ticking from a clock, features on a device or in a piece of software or a colleague talking too loudly; even if something isn't actually damaging you, if you don't want it, it's a harm. Harms are great because they provide an opportunity for you to improve your system.

For more on Ideality, pop over to Chapter 5.

From perfection to reality: Defining the Ideality you want

When thinking about the systems you're creating, considering the Ideality you want is useful. Creating a wish list of benefits that you want in an ideal world is all very well but you don't live in an ideal world. When you take a step back towards reality, you can start identifying how many of the benefits you'd be happy with, and the costs and harms you're willing to tolerate.

Try ranking your benefits into 'must haves' and 'nice to haves' to help you develop a realistic picture of your tolerances before you convert the benefits into practical requirements. Check out the example in Table 9-3.

Table 9-3	Specifying Your Ideality
Ideal Benefit	*Requirement*
Must haves:	
Can carry an infinite amount of water	Carries 500 ml–3 litres
Always clean (inside and out)	Easy to wash or dishwasher-proof or self-cleaning
Easily portable (zero weight, zero volume)	Light
	Small when empty
Nice to haves:	
Can carry any liquid (squash, coffee, fizzy water)	Can carry any room-temperature liquid
Keeps water cold	Same

You can then consider the costs and harms you're willing to put up with, as shown in Table 9-4.

Table 9-4	Balancing the Costs and Harms
Costs	*Harms*
Costs less than £1 to produce	Disposal may damage the environment; consider recycling
1 year to develop, design and test	Can survive being dropped from 2m
Sells at no more than £5	

Establishing Ideality can be very useful at the start of a problem-solving process as a method for ranking and selecting solutions at the end; it also means that everyone understands what they're aiming to achieve without getting stuck in solutions. A civil engineering firm in New Zealand starts every project by all stakeholders and interested parties defining the project's Ideality and states that further TRIZ is unnecessary because Ideality defines its scope so well that how to achieve its objectives is immediately obvious.

Learning to compare apples and pears

One of the encouraging things about developing an Ideality audit is that it can help you to efficiently compare very different solutions. You can create a table of the benefits, costs and harms and score any systems/solutions against the Ideality you've specified.

If you want to get detailed, you can attribute numbers and weightings to your rankings, multiply them, add up all the benefits and divide them by the costs and harms. I caution against that approach, however, because you end up with a very precise number – and precise isn't the same as accurate. Both the rankings and weightings will be informed by the knowledge and assumptions of the people in the room. Lots of guesstimating will occur, and if you have one solution that's given a score of 7.2 versus another that's scored 7.5, it really doesn't mean very much if the rankings and weightings haven't been properly researched.

That said, your Ideality does provide a useful framework for comparing systems in a relatively objective and rational way. It's particularly useful when you're choosing between options that already exist, such as buying a house, car or computer. And it will also help you compare very different systems, for example a car versus a motorbike, a computer versus a tablet or an old house with lots of charm versus a new house that requires very little upkeep.

Your Ideality is also a good way to identify what costs you're willing to put in – both upfront and over the life of the system. Many quick fixes actually end up costing more in the long run, and considering the Ideality of time, money and effort will help you uncover such short-term solutions and make the best decision for the future. A huge budget was allocated to building sewers in London in the 1860s, which has resulted in much lower running costs in terms of time, effort and money over its lifetime. Conversely, many public bodies in the UK are required to accept the lowest tender for projects, which may ultimately cost far more in the long run. Sensible consideration of your Ideality (in 9 Boxes) will help you make good decisions with all the necessary information.

Thinking about the unthinkable: Managing risk

Considering your Ideality also helps you to think about risk – which most people don't enjoy focusing on. Risk refers to unintended downsides, things that may go wrong, and it's human nature to avoid dwelling on uncertainty. It's one of the reasons why people find writing a will or choosing a life insurance package difficult.

Sometimes you can also be so excited about a particular solution that you ignore or are blind to the downsides. Your focus on value (benefits over costs) may completely ignore potential downsides (particularly sustainability and environmental issues). Someone responsible for implementing new service products for his company once told me that Ideality had revolutionised his job. He used to compare and rank ideas based on value, but having discovered TRIZ, he added in 'risk' as another factor. He then considered not only how much time and money would be involved in implementing a solution and how much revenue it would generate, but also what would happen if it went wrong; such as the effect on customer loyalty, for example.

The TRIZ extreme thinking in terms of the Ideal Outcome – getting out your magic wand, releasing your inner Spice Girl ('I'll tell you what I want – what I really, really want') and learning to be appropriately demanding – thus starts off wacky and ends up rooted in the practical. Like so many of the TRIZ tools, it takes you out of the real world and into the realm of the possible. You then have a number of systematic steps to follow that show you how to turn that Ideal Outcome into reality. By proposing extreme solutions and understanding what you want, you may well end up with something truly innovative and groundbreaking.

Chapter 10

Problem Solving and Being Creative with Others

. . . .

In This Chapter

▶ Pushing the boundaries of your thinking

▶ Developing problem-solving persistence

▶ Working creatively with other people

. . . .

*B*eing a genius at your job and a whizz with TRIZ only gets you so far. The psychological and social aspects to problem solving can make a huge difference to its success, and the effective application of TRIZ also helps you manage these aspects effectively.

This chapter considers how best to use TRIZ when working with others, to ensure you come up with the best results and experience the least pain along the way.

Going for What You Really Want

One of the most important philosophical aspects of TRIZ is that it helps us believe that things can be better. You start by saying that it's possible to improve anything and, even more boldly, that you'll get everything you want. This positive attitude takes you much further than starting with a pragmatic attitude of trying to find the easiest, smallest change possible.

If you need a solution that's easy to do, then that can be part of what you're looking for, but with TRIZ you start by listing all the things you want and trying to go for as many as possible.

Approaching a problem with a negative attitude – thinking 'this is impossible' – will doom you to failure, and the fact that you're tackling an issue with TRIZ should give you confidence that you'll find good, and possibly the best, solutions. I've been involved in many urgent, critical and therefore scary problem-solving sessions where a component has failed, something has started leaking or a department is facing closure if a solution isn't found in a matter of days, for example.

While it's good for everyone involved in problem solving to understand the importance of working on an issue, starting a problem-solving session in a state of terror isn't helpful. Understanding that they're following a logical, systematic problem-solving process (see Chapter 11), which, if followed, generates deep understanding and clever solutions, will help a team have confidence – if not in themselves, then in the process.

The fundamental attitude throughout the whole process should focus on what you really want, even if downsides exist, rather than on only what you think is achievable. This is one of the ways in which the Ideal Outcome (find out more about this in Chapter 9) is so important: you start early on in the process by considering what you want, and listing out all requirements, whether you think they can be achieved or not. Because the Ideal Outcome is by definition a stretch goal (it's ideal – therefore you're unlikely to actually achieve it), it allows what people want to come through more easily than trying to define realistic requirements. You don't have to have a debate about cost, for example, because in an ideal world your solution doesn't cost anything.

Often when problem solving, people don't know where or how they could tackle the problem, and they feel really stuck. Understanding what you really, really want helps you get out of the detail and the current way of doing things and identify the direction you need to be travelling in to find solutions. When you've understood what you want, you can see all the problems that currently exist and then tackle them one by one.

Starting by striving to achieve all the things you want ensures that you identify all of them (in your Ideal Outcome), and that attitude makes you lift your head and look to the horizon. A client once told me that completing the Ideal Outcome is one of the most fundamental steps in innovation for him because, as a result of starting with looking for what they really want, his team uncovers more requirements than they would using traditional problem-solving methods. Setting a distant goal means they travel further towards the Ideal than they would if they set a more pragmatic, realistic one. They sometimes surprise themselves with how far they travel and how innovative the resulting new products are.

Another part of the TRIZ philosophy is that you'll get everything you want without changing anything. As soon as you've identified what you really want, you can see how to achieve it using what you already have, with sensible use of your available resources.

Say you want to put up a sign warning people when the road is icy. Ideally, the sign would only be apparent when the road is icy and at no other time. One clever invention, shown in Figure 10-1, uses the cause of the ice – the cold – to change the colour of the sign to warn of potentially icy roads. This is a very clever, resourceful invention, and a great example of a *self-system*: no power or human intervention is required; instead, the cause of the problem automatically provides the solution.

 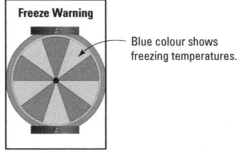

Blue colour shows
freezing temperatures.

Figure 10-1:
A clever
self-system.

Illustration by John Wiley & Sons Ltd.

Going for what you want should become a mental reflex for seasoned TRIZniks. Whenever you're trying to work on something, choose something or even identify what it is you want, and then ask yourself what you'd get in an ideal world or if you had a magic wand. You're brought up to focus on the practical – to be realistic. Formal education reinforces this message. However, when you're a child you think everything is possible. At 5 years old you think you can eat three ice creams in a row and nothing will go wrong; when you're older and wiser, you know that over-indulging on ice cream will make you feel sick. You need to recapture that 5-year-old child's imagination and not squash that longing for three ice creams. You understand that's what you want, then work out how you can achieve it without incurring the downsides: maybe you can have three, small, differently flavoured scoops.

When you've established what you want, you can then identify the problems associated with getting it – and tackle those problems with TRIZ, logically and systematically. It's important to bear this in mind: looking for what you want without worrying about downsides isn't some fluffy tree-hugging approach to creativity, just unlocking your brain and thinking positive thoughts! Instead it requires you to explore all requirements logically and rigorously and then deal with any inevitable problems. You don't let the inevitable problems bog you down too early in proceedings; there's a time and a place for thinking of constraints, but in TRIZ you do it second – not first.

This approach can make problem solving go much faster. You've probably experienced meetings in which someone suggests an idea and then everyone spends half an hour discussing its problems and downsides. What I call 'ugly baby wars' are deeply boring (the ugly baby is a beloved solution – the person who generated it loves it and thinks it's perfect and everyone else thinks it's hideous). Debating the pros and cons can be exhausting and is a bit like falling down a rabbit hole: you're in the detail really early on and lose sight of the big picture. Often the most practical or easiest to implement solution can seem the most appealing, but it may be a short-sighted decision.

Starting with what you want pushes you towards innovative solutions, and knowing that you can deal with any of the inevitable problems by tackling them with the TRIZ problem-solving tools gives you confidence. You *can* solve contradictions, deal with harms, improve insufficiencies and so on.

Thinking in Extremes

Another important part of the TRIZ philosophy is learning to stretch your thinking beyond the practical. Pragmatic thinking is important and useful: being able to focus your attention within reasonable boundaries is vital for efficient thinking in your daily life. However, if you can learn to look beyond these practical boundaries, you will find innovative ways of looking at your problem which will lead to creative new solutions.

Challenging constraints: Real or imagined?

One of the major benefits of stretching your thinking beyond practical limits is that it helps you challenge constraints, and by challenging them you discover whether they're real or not. You may find that they are real, but it's still worth asking the question. Trying to find solutions within imaginary constraints makes your job needlessly difficult, and constrains not only the kind of solutions you come up with but also your thinking, inhibiting the flow of ideas.

The *nine dots* problem demonstrates how false constraints inhibit clever thinking, and solving it is often given as an example of insight problem solving and creative thinking. Participants are shown nine dots, as in Figure 10-2, and asked to draw five continuous, straight lines, which pass through each of the nine dots, without their pencil leaving the paper. This is easy: they usually come up with a solution, as shown in Figure 10-3. Then – and this is the hard bit – they are asked to complete the same task, but using only four continuous, straight lines. While this appears to be an easy problem in the sense that lots of potential solutions exist, in fact it's psychologically very difficult; generally, fewer than 5 per cent of people see the solution. Why is it hard? Because people construct a false constraint: they see a 'box', created by the outer dots and reinforced by the solution they found when drawing five lines, and are reluctant to draw outside it. No box exists for this second problem – only nine dots – and even if it did, no one's telling them they can't draw outside it! In fact, even when people have been told explicitly to 'draw outside the box', they often struggle to find the solution (shown in Figure 10-4). Not until they see a drawn hint can they find the answer.

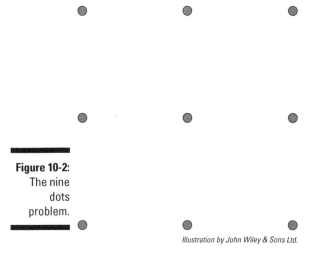

Figure 10-2:
The nine dots problem.

Illustration by John Wiley & Sons Ltd.

This problem has been researched extensively by psychologists over the years and the findings have been repeated over and over again. What it reveals is that thinking 'outside the box' is very difficult without prompting. You, like most people, need explicit guidance to reconstruct the conditions of your problems and check constraints. You can't trust that you're seeing things correctly, as you may be subject to a range of unconscious biases and assumptions – sometimes based on previous problem-solving success – that form a kind of psychological inertia.

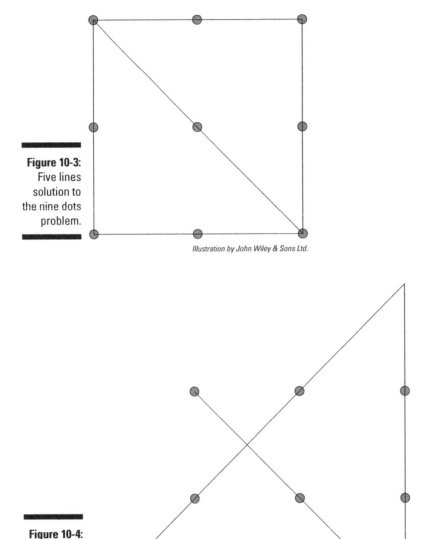

Illustration by John Wiley & Sons Ltd.

Figure 10-3:
Five lines
solution to
the nine dots
problem.

Figure 10-4:
Four lines
solution to
the nine dots
problem.

Illustration by John Wiley & Sons Ltd.

Psychological inertia is a mental rut. When you experience *psychological inertia*, your thinking is blinkered by unconscious assumptions that hamper your ability to see things as they really are. You can challenge your thinking and ensure you're not stuck in a rut by restructuring your view of the problem.

Doing so helps you challenge your assumptions about the nature of your problem and the kind of solutions that are acceptable. (Check out Chapter 7 for more on tackling psychological inertia.)

Restructuring problems and solutions

All of the TRIZ tools help you restructure how you see your problem and possible solutions. Very often you'll find that what you considered a constraint isn't real, and for this reason problem solving with other departments is essential. When you ask all relevant people involved in the development and delivery of solutions to work with you (from marketing to the supply chain), you often find that what they're willing to accept is much broader than you'd assumed.

One common error when problem solving is working on someone's 'bad solution' to an underlying and more fundamental problem. When you complete the Ideal Outcome (see Chapter 9 for more details), you decide at what level to scope your problem and what type of solutions to look at. You establish whether you want to keep improving that initial bad solution or go up a level and look at the more fundamental problem this solution is tackling.

I facilitated a TRIZ problem-solving session focused on improving a complicated bit of kit for an underwater energy-generation device. Only when the team completed their Ideal Outcome did they realise that the company was using this device because it wanted faster installation – and the device was only one way to achieve this outcome. They then took the scope of the problem solving up a level and looked at ways of getting what they really wanted: faster installation. The team generated lots of clever, easier solutions to the problem as a result of changing the scope of the problem.

Another powerful tool for restructuring your view of a problem is Thinking in Time and Scale (explained in Chapter 8). This helps you understand the context and detail of your problem and how they relate, which may result in seeing opportunities that have eluded you in the past. Numerous problem-solving sessions have been cracked open by looking at the impact of time and scale as people realise that many other opportunities exist beyond their current focus.

Thinking crazily (while being practical)

A large part of the TRIZ magic comes from the ability to think wildly and then turn those wild thoughts into reality. Pragmatic and sensible behaviour is what's usually rewarded and appreciated at work. If a colleague complains that the printer's jammed, she probably wants your help to pull out the mangled paper, rather than a suggestion that she go paperless, invent a new printer or communicate entirely through the medium of interpretive dance. However, when you need creative thinking, you have to step away from that pragmatic, sensible approach and be open to all ideas.

Don't get me wrong – TRIZ is still very serious and will end up delivering practical solutions because TRIZ thinking is both very imaginative and entirely grounded in reality. TRIZ takes you on a little holiday from your normal pragmatic approach – in the sense that it's a place where you can have fun and play – but you don't stay there; ultimately, you come back home. This process allows you to think impractically as your brain runs free, but then to work out how to turn those impractical ideas into a practical solution.

Very serious technical people sometimes resist this approach because they worry that generating impractical solutions will be a waste of time, and nothing useful will come out of it. However, there's something important and serious about thinking more freely, outside the boundaries of reality. It's one way in which to challenge your psychological inertia and think of more ideal solutions. The TRIZ creativity tools help you to think in a different way, primarily Smart Little People and Size–Time–Cost (both Chapter 7) and defining your Ideal Outcome (Chapter 9).

The TRIZ solution tools, based on patent analysis – the 40 Inventive Principles (Chapter 3), Trends of Technical Evolution (Chapter 4), Standard Solutions (Chapter 13) and Effects Database (Chapter 6) – also stimulate your creativity but in a very focused way, around conceptual solutions that have been used successfully in the past.

If some of the suggested solutions seem improbable, don't give up. Inventive Principle 13, for example, suggests doing something the other way round or upside down, which may seem very wacky but has many simple and practical applications that are obvious in retrospect. Putting the opening for ketchup bottles on the bottom, not the top, means gravity aids the passage of the thick and viscous liquid. This solution has been repeated in other places; most hair conditioner bottles now also have their opening on the bottom, which also makes them easy to distinguish from the right way up but otherwise matching shampoo bottle even with closed eyes in a shower.

Because TRIZ provides a structured process for problem solving (tackled in Chapter 11), people who may otherwise be resistant to this very free thinking can relax: their thinking is both highly structured and utterly free. TRIZ solves this contradiction by providing individual periods of very free thinking within a series of structured steps.

Being Persistent in the Face of Failure

Successfully creative problem solvers are known for their persistence. TRIZ gives you not only the confidence to be persistent but also a practical toolkit for helping you drive solutions forward and find good solutions faster.

Let's look at how this toolkit works in the following sections.

Improving imperfect solutions

TRIZ is fantastically pragmatic when it comes to solutions, to the extent of being really quite strict. You're always striving for the ideal, and any solution you come up with is likely to be imperfect. As a TRIZ problem solver you want to find all the imperfections in your solutions so that you can improve them. This is one reason why capturing everything you want is so important: any place that your existing system or solution doesn't meet your Ideal is an opportunity for improvement.

When you have a well-defined system or solution already in practice and you know how it works, you can map all its problems using Function Analysis (explained in Chapter 12). This pulls out any harms or insufficient actions that you can improve using the Standard Solutions and any contradictions you can solve using the 40 Principles (flick back to Chapter 3 to find out more about these).

A quick route for improving solutions – particularly if they're new and thus not well defined – is to identify any contradictions. Take your solution and identify what's good and bad about it in a simple table such as the one shown in Table 10-1. When you've identified what's good and bad in normal language, you can translate that into Technical or Physical Contradictions, and look up which of the 40 Inventive Principles will be able to improve your system.

The example shown in Table 10-1 is taken from a recent problem-solving session with one of Highways England's maintenance suppliers: together these organisations manage a high volume, complex programme of highway engineering schemes on an annual basis. The majority of the works are planned, but unplanned maintenance repairs are also needed as issues arise on the highway network.

Currently, not one single overarching programme exists to encompass all the construction schemes (both major and minor) plus smaller maintenance activities that are taking place on the highway network in a maintenance area.

Table 10-1			Problem-Solving Table		
What Do I Want?	*Bad Solutions*	*What's Good?*	*What's Bad?*	*TRIZ Solution Trigger*	*Better Solution*
Better planning	Fixed agreed programme	Everyone knows what's happening when	Difficult to react to unplanned events	Inventive Principle 35. Parameter Change	Plan and fix big projects first. Then increase the flexibility of planning as the projects decrease in size and complexity
		Better resource management	Not enough info for fixing dates		
		Cost certainty	Miss oppor-tunities		
		Parameter 13: Stability of the object's composition	Parameter 35: Adaptability or versatility		

The current programme that exists for undertaking maintenance activities can be quite volatile and results in reactive working, and this programme does not take account of major project activities on the network. Coupled with the fact that the annual investment in the roads network will double in the next two years, this leads to more pressure on the delivery of an overall stable programme of works.

The team concluded that an ideal works programme (seeking Ideality) would be one that encompassed everything and was totally rigid and was always adhered to. However, it was understood that this was not practically achievable due to unplanned reactionary activities. This therefore led to the view that we needed a programme that was both rigid and flexible, which is the contradiction: how can you have a programme that is both rigid and flexible?

Looking up this contradiction in the Technical Contradiction Matrix (see Chapter 3) suggested Principle 35, Parameter Change. This principle suggests establishing varying degrees of flexibility for the different projects – casting large, complex projects in stone and increasing flexibility the smaller projects become. This was a revolutionary moment for the team, as the suggestion was both highly innovative for the company and relatively easy to implement.

Another approach for improving solutions is to conduct an Ideality Audit of your solutions. *Ideality* is a measure of how good something is, as explained by this funky equation:

$$Ideality = \frac{\sum benefits}{\sum costs + harms}$$

Good, eh?

This identifies all the benefits your solution is delivering, and where those benefits may be insufficient (an opportunity to apply the Standard Solutions in Chapter 13). If you have missing benefits, you can apply X-Factor thinking and the Effects Database to deliver them (both Chapter 6). You identify all costs and also how you could potentially reduce or remove them by trimming out components while keeping their useful action (see Chapter 14). Any harms you identify, you can deal with by applying the Standard Solutions for harms (described in Chapter 13). You can also take any solution or system and improve it by driving it up the Trends (see Chapter 4). Job done!

Fixing things when something's gone wrong

Don't give up! You've just hit a problem. And you love problems – you're a TRIZ problem solver.

TRIZ problem solving is an *iterative process*, which means you don't just do it once: you get the best results when you repeat the process on any solutions you generate, as each time you do so you get closer to your goal. As a result, TRIZ is often a suggested approach during project development, for example, as part of an already established process. Much of TRIZ's focus is on the generation of new solutions and new methods of working at the first stage of a project. This is a good approach, because it's best to come up with a strong solution early on rather than fix up weak solutions.

However, the reality of life is that even the best solutions run into problems: for this reason, TRIZ is useful during any stage of a project. Being surprised when this happens is pointless; rather, you need to plan for it. I worked with a team of engineers who planned to use TRIZ together with a stage-gate process: their product development process was separated into a number of stages, each of which had a 'gate' where decisions were made about whether (and how) to go forward. At any point in the process when the product hit a snag or something went wrong – a problem with the concept, the design, the prototype, manufacturing, the supply chain – the team would apply TRIZ.

TRIZ is also a useful addition to any process that involves reviews, such as design reviews: any problems or shortcomings that have been identified can benefit from the application of TRIZ.

The point is that things do go wrong, but now you know what to do next! Use TRIZ!

Knowing when to stop

Sometimes, despite your best efforts, the solution you've been working on just keeps going wrong, and for every step forward you seem to take at least another one back.

While most problems can be addressed by applying TRIZ, in some cases TRIZ won't be able to help:

- **When emotional, political or any other non-logical reasons are preventing you moving forward:** TRIZ is logical and assumes you are too. If your solution is being hampered by a political fight between your boss and that of another department, you'll need to deal with that issue separately.

- **If you can no longer change things:** If the outcome has already been decided or resources have disappeared (for example, budget, time, people), unless you can reframe the issue to include managing these issues, you'll just have to accept that you can't do any more.

- **When you're the only one who cares about solving the problem but the solution relies on other people:** You need to engage them before you apply TRIZ or you'll keep running into a brick wall.

- **If you lack the knowledge to make something work:** Fortunately, you can usually find and engage that knowledge, and I recommend calling on the help of experts when needed (academics are often helpful).

Sometimes you're held back by circumstances beyond your control. Luckily for you, there are usually ways around problems, and if you keep a record of your problem-solving process, you can go back a step or two and find alternative routes forward.

You can take whatever's getting in your way and frame it as a TRIZ problem to be solved (whether it's your boss's personality or lack of resources or any of the other problems suggested in the bullet list above). What's important is that you tackle them separately to the issue at hand.

Sharing and Developing Ideas with Other People

Playing nicely with others is a skill that extends beyond the classroom and into the boardroom and beyond. The ability to share and develop ideas with other people can be tough, but is important. Different people will help you look at problems from different perspectives, which is essential for good problem understanding; having a variety of expertise and knowledge also broadens the scope of potential solutions you will be able to generate. Diversity of thinking is vital for creativity, and your colleagues can inspire and motivate you, as long as the diversity is both embraced and well managed: using the TRIZ process and attitude to problem solving helps make this possible.

Recognising why bad solutions are a good idea

The phrase 'bad solutions' doesn't appear anywhere in the original Russian TRIZ literature (as far as I'm aware), but it's a crucial part of the TRIZ approach. *Bad solutions* tell you both something about what you want and what's getting in the way of your achieving it. This knowledge gives you a very direct route to generating more and better solutions.

The most famous TRIZ tool is probably solving contradictions. A contradiction arises when you want something but it has a downside – in an actual, practical solution. In that particular solution you can't get one without the other, but this 'bad solution' gives you incredibly useful information both about what you want and what is stopping you getting it.

Going back to our cheese grater example in Chapter 3 (if you've not yet read that chapter, we wanted a cheese grater that grates a lot of cheese but doesn't make a mess), no contradiction is evident in these desires. The contradiction exists in a specific solution: that as we make our cheese grater wider and thus have a larger surface area, so we get faster grating, but the cheese then goes everywhere. Lots of other solutions may give us what we want without this contradiction, and we can find and access them by starting with just one bad solution. We take our initial bad solution and understand what's good and bad about it, uncover and then solve the contradictions, and generate new, better solutions. Often these solutions look very different from our initial solution. This is another example of TRIZ being both very conceptual and also rooted in the nitty-gritty of real-world problems. You can start with a specific solution, understand what's good and bad about it, uncover all the things you want, find out what's missing or imperfect in that solution and come up with something much, much better. And this is why bad solutions are so important: each one tells you something interesting and important about what you're looking for.

Don't be afraid of your initial, top-of-the-head ideas, as they can provide useful information. Consider starting a TRIZ problem-solving session by brainstorming because it can help you begin to think of all the things you want in advance of completing the Ideal Outcome (described in Chapter 9). Capturing all solutions allows people to focus on the process rather than their preconceived outcome. People attend TRIZ sessions armed with their own solutions and are unable to concentrate fully until they've shared them. They won't be able to help themselves because they love those solutions.

Seeing the flaws in your own solutions

All TRIZ processes are about taking bad solutions and making them better. You've got to learn to let this happen to your own solutions too: accept that you love them but that as they go through the TRIZ process they may change beyond recognition. Don't hold them back – they're achieving their destiny.

You can't help but love your own ideas. Like ugly babies, you think they're beautiful and perfect and can't see anything wrong with them. Your colleagues, however, can see every flaw that you're blind to. This situation is okay. You need to love your ideas if you're going to have the motivation and stamina to develop them into good solutions and put them into practice. But it's also important to understand that a problem may be solved in many, many ways, and your solution is just one way. Your colleagues will also have their own solutions, and you'll be able to spot their problems. You need to regard each other's views as a useful resource and an inevitable side effect of the creative process. It's much easier to love your own solutions than someone else's, but you have to learn to share your own solutions with humility and listen to other people's with good grace if you're going to work well in a team.

Ideas versus solutions

Many people use the words 'ideas' and 'solutions' interchangeably. In a practical sense, the difference matters very little. However, the reason for saying *solution* is that the thought you've had is a solution to a problem or meets a need of some kind: it's goal-orientated. An *idea* doesn't need to solve a problem you've uncovered; rather, it can be an unformed thought or a suggestion about another way of doing things that doesn't initially seem to be relevant to the problem at hand.

An idea can also solve a problem – it can be a solution – but it may not be. A *solution* is a more precise term for a thought that's directly solving a problem; an *idea* is a more nebulous, top-of-the-head thought. When you're running a problem-solving session, you can treat all ideas as solutions because you're tackling a problem.

Learning to love other people's solutions

By calling your ideas 'bad solutions,' you lower the threshold for sharing. Your idea doesn't have to be good – it can actually be terrible. But because you're prefacing it with 'bad', you're acknowledging it has faults and don't have to feel embarrassed about it. When you try to generate solutions, you'll always come up with a whole range of ideas, from the well-defined and brilliant to the interesting-but-flawed and wacky, impractical and really off the wall. The wilder solutions will probably have bigger problems, but your brain will have generated them for a reason. You need to capture all those solutions because ways may exist to deal with the problems. If you don't feel judged, you'll share more ideas.

One of the most important aspects of innovative teams is trust: people feel comfortable sharing their solutions with colleagues because they know they won't be laughed at or ridiculed. Any idea, no matter how wacky, impractical, improbable and bizarre, tells you something interesting about what you're looking for and may have a seed of genius buried within it. By creating a culture in which all initial solutions are termed 'bad solutions' (that is, they're imperfect and can be improved upon), everyone will find it easy to share ideas with others and build upon them.

Capturing, sharing and combining solutions

When you share everything you're thinking, your brain is running free and as soon as you've had – and let go of – the first couple of solutions, you'll generate more, usually better, ideas. You need to generate flow for effective creative thinking, and not stopping to judge or critique ideas keeps it going. Bad solutions can also stimulate someone else's creativity: she may see a problem with your solution and know how to fix it, or see another way of putting the concept behind your idea into practice.

Capturing all solutions in a *Bad Solution Park* means you can share them very effectively. One method is to go around the room asking everyone for their solutions, one at a time. This approach ensures that everyone contributes at least one solution, which is shared and captured. Some people find it easy to generate many solutions to a problem very quickly, and they'll have a huge pile of sticky notes in front of them; others may generate only one or two, and require more time.

Using a Bad Solution Park and asking everyone for one solution at a time also ensures that everyone's voice is heard equally to begin with, and not just the person with the loudest voice or most solutions. Some people are more reserved and can feel intimidated by the thought of talking about their solutions in front of a large group of people. Taking this approach acknowledges that everyone can and did contribute solutions.

Put a Bad Solution Park on the wall during a problem-solving session and review it from time to time. People can slip their ideas on it as they come to them, which benefits those who are shy. Problem solving isn't a competition: you want to find the best solution to the problem at hand. A Bad Solution Park ensures that everyone can contribute in the way most comfortable for him, and that everyone's solutions have been shared and can be developed and built upon.

Running a session explicitly for combining and hybridising solutions is a good idea. Often the best solutions emerge from the combination of a number of solutions rather than the idea of one person. Combining ideas creates a hybrid solution of the best bits of individual ideas. Dog breeders create hybrid breeds in this way (for example, the Labradoodle – a cross between a Labrador Retriever and a Poodle), and you too can identify the plus points of each solution and combine them to create your very own Labradoodle solution.

Hybridisation is also known as creating a 'cabbage-carrot'. Cabbages have delicious leaves but the roots aren't useful to us – they take up space, use water and remove minerals from the soil. In contrast, carrots have delicious roots but useless leaves. A hybrid would have delicious and useful roots and leaves (alternatively, we could plant beetroot instead).

Creating a climate for innovation

Innovation is the successful and practical application of creativity, taking novel ideas and putting them into practice. For organisations to be innovative, they must encourage staff to find opportunities for innovation, generate creative ideas and implement them.

The most successfully innovative organisations create a conducive climate by:

- ✔ **Providing resources:** Offer practical support in the form of money and time (scheduling projects to include time to try new things).

- ✔ **Supporting and encouraging creative work and innovative thinking:** Praise effort rather than outcome – including innovation and problem solving in performance appraisals.

✔ **Allowing people to take risks:** Don't punish them if something goes wrong.

✔ **Giving people autonomy:** Allow people to tackle their work as they see fit without micro-managing them.

✔ **Encouraging sharing of information:** Support collaborative, cross-functional working across the organisation.

✔ **Promoting diversity:** Create teams of people from different backgrounds, with different experiences and approaches to thinking to generate more creative thinking.

✔ **Communicating a clear vision:** Ensure clear goals and strategies for achieving them are developed and communicated at both the organisational and individual level.

✔ **Allowing focus:** Give people space and time to concentrate on the most important tasks.

✔ **Matching people with the right skills and experiences with the right tasks:** Motivate and stretch them by allowing them to build on their existing skill set.

✔ **Providing training:** Offer courses in creative thinking skills – like TRIZ!

If creating an innovative organisation is beyond the scope of your control, you can still encourage innovative problem solving and creative thinking in yourself and the people around you by:

✔ **Being positive:** Believe it's possible to change things and see what's good in other people's suggestions.

✔ **Suspending judgement:** Don't judge ideas or new thinking too soon; give yourself some space to play and explore new approaches.

✔ **Being curious:** Look for problems and places to improve things.

✔ **Developing your skills:** Enhance your ability to put new solutions into practice and find new ways of doing things by broadening your professional experience and practising your creative thinking skills whenever possible.

✔ **Trusting:** Create a supportive and positive environment that encourages people to listen to each other and share ideas.

✔ **Building on each other's ideas:** Don't reinvent the wheel!

✔ **Staying motivated:** If you work in an area you find interesting and satisfying, you'll be more creative.

✔ **Challenging constraints:** Always question what's possible.

✔ **Building a community:** There'll be other people around you who are interested in being more innovative. Get together and help each other on projects – you'll get a chance to practise your creative thinking skills and learn about other parts of the organisation.

✔ **Scheduling time for creative thinking:** Allow time for generating new ideas, trying them out and dealing with any problems that may arise.

✔ **Having fun and remaining serious:** Enjoying yourself helps you think more creatively; understanding that things are serious helps you find more problems with the way things are – both of these make it easier to stay motivated, keep working and remain engaged with the process.

Part IV

Understanding, Defining and Solving Difficult Problems with TRIZ

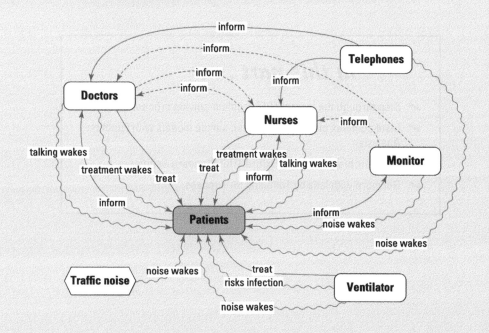

Visit www.dummies.com/extras/triz for a free article that gives an example of how Thinking in Time and Scale can make impossibly complex issues simple to understand.

In this part . . .

✔ Step through the logical TRIZ problem-solving process.

✔ Distil complex problems into clear, simple models with Function Analysis.

✔ Solve problems with the Oxford TRIZ Standard Solutions.

✔ Get more with less by Trimming for success.

Chapter 11

Applying the TRIZ Problem-Solving Process

··

In This Chapter

▶ Thinking creatively by following the rules

▶ Following the TRIZ problem-solving process step by step

▶ Working as a team

··

Clever problem solving requires clear and systematic thinking: the TRIZ step-by-step process explored in this chapter can help you work through your problem in a logical way to help you find breakthrough, innovative solutions.

The steps, in a nutshell, are as follows:

1. **Understand and scope the problem.**

2. **Uncover all needs and scope the solutions.**

3. **Zoom in and define the problem.**

4. **Identify the solution triggers.**

5. **Generate solutions.**

6. **Rank solutions and implement.**

This chapter puts some factual filling into this theoretical sandwich.

Logically and Systematically Solving Problems

When faced with a difficult problem, what do you do? Go for a walk and hope a solution will come to you? The TRIZ approach is to work systematically to understand and then define and focus your problem, in order to find the best solution.

The more solutions you generate for the wrong problem, the farther away they take you from where you want to be. When you have a difficult and confusing problem, make sure you have a process that you can follow – particularly when you're working in a team.

Different people have different approaches to problems, and preferred ways of tackling them. Some like to undertake a brief problem-understanding process and jump to brainstorming as soon as possible – fast but risky; others prefer to spend a lot of time rigorously working through and understanding the problem before thinking of solutions – safe but slow. Getting people with these very different approaches to work together can be difficult, which is where the Oxford TRIZ problem-solving processes are so powerful: while they're logical and systematic, they're also fast and flexible. Understanding a problem doesn't take a week; it can be done in two to four hours by a team, or even less time if you're working alone. TRIZ is rigorous enough to be embraced by people who like making sure the problem has been fully understood, but fast enough to keep the brainstormers happy. The *Oxford TRIZ processes* bottle all the logic of ARIZ (Algoritm Resheniya Izobretatelskikh Zadatch, or the Algorithm of Inventive Problem Solving) into simpler, more flexible processes.

ARIZ is a logical, long and rigorous approach to solving problems, and was Genrich Altshuller's (you can find more on him in Chapter 2) attempt to provide a single but many-step process that could be used to tackle any problem – no matter how complex and difficult. As a result, it's fantastically powerful for the most difficult problems but can feel like overkill for simpler problems – and daunting and off-putting to novice TRIZ users. It also has one critical and difficult part – choosing the right fundamental contradiction – which is essential for success. One wrong choice at this point and all the careful hard work over the many stages is wasted. By contrast, the Oxford TRIZ processes are simpler, faster, less risky and more flexible. They're easier to pick up and use but, like any new skill (such as learning a language or playing a musical instrument), expertise in a new way of tackling problems takes practice and experience; you can't develop confidence and speed overnight. In this chapter I explain the Oxford TRIZ process.

Trusting the process (and yourself)

One of the benefits of following a trusted problem-solving process is that it allows you to focus your attention and energy on the problem, and not on the process itself. You don't need to work out how you should be looking at the problem – you just follow the steps. To do so, however, it's necessary to trust the process. This is where TRIZ's engineering and scientific pedigree comes in useful. It isn't a management tool that's been dreamt up by someone clever on a beach: it's the logic of engineering problem solving distilled into a systematic process. Although it comes from engineering and science, this approach is valuable for any kind of problem – management problems, business problems, even personal problems – because it provides clarity of thought, clear steps and a fresh approach. This can inspire confidence when you have difficult problems to solve and desperately need answers; you know that when you follow the logical steps you'll find an answer.

When you trust the process, you can trust your own thinking and the ideas that you generate at every stage of the process. This frees up valuable mental space for problem solving and clever thinking, and gives you confidence – in the method, and yourself.

Making big leaps by taking small steps

I've noticed when working with teams that very few people find it really easy to make big leaps when thinking of solutions. Those who can often think of very interesting, left-field solutions, but leave the rest of the team scratching their heads thinking, 'How did she get there from here?'

One of these left-field thinkers once told me that TRIZ helped him show everyone else why his thinking went to that far-off place: TRIZ would always be able to explain (and predict) the kind of leaps his mind made anyway, and as a result, TRIZ helped him work more effectively with others. TRIZ would help him explain why he went in that direction, but provided a series of stepping stones for everyone else to get there too. The team he was working with could then generate a range of solutions, from minor tweaks to increasingly radical changes that led all the way to his big mental leap. Following this, the team could work around developing and improving his initial solution, rather than rejecting it as too wacky and off-the-wall, and having to be persuaded by him to take it on board.

But more excitingly for you (and everyone else), TRIZ helps *you* cover the ground made during those big leaps by others. You can get to very innovative solutions by following the logical processes, but instead of having to make one big jump, you follow many smaller steps and end up in the same place. Initially, you'll generate incremental improvements to your system, but then, as you become confident in taking those small steps in the TRIZ problem-solving process, you end up being able to cover the ground made by those big leaps yourself – reliably.

Developing your own problem-solving map

I've used the process described here for over ten years on many different kinds of problems in all kinds of industries – and it always delivers results. But the reality of using TRIZ is that you always have more than one choice of tool to use and, depending on the problem at hand, may want to apply them in different orders. For example, for focused problems you may want to do a simple Ideal Outcome before mapping the context of the problem in

9 Boxes; you might even go straight to Function Analysis. For issues where part of the goal is to gain team consensus regarding the scope of the problem, it's more useful to do the context mapping first, as this can help define the correct scope for the Ideal Outcome as well as get the team onboard. The most important thing is to make sure you cover the basic processes described in this chapter.

If you want to develop your own problem-solving map, make sure it includes the five fundamental stages shown in Table 11-1.

Table 11-1   Which Tools to Use at Each Stage of Problem Solving

Problem-Solving Stage *Every Stage of the Process*	Most Useful TRIZ Tools *Bad Solution Park*
Understand and scope the problem	9 Boxes context map, hazards or causes of problems maps (Chapter 8)
	Ideality Audit (Chapter 5)
Uncover all needs and scope the solution	Ideal Outcome (Chapter 9)
	Ideal System (Chapter 9)
	9 Boxes (Chapter 8)
Define the problem	Function Analysis (Chapter 12)
	Technical Contradiction Matrix (Chapter 3)
	Separation Principles for solving Physical Contradictions (Chapter 3)
	Smart Little People (Chapter 7)
Identify the solution triggers	Standard Solutions (Chapter 13)
Generate solutions to the problem	Trimming (Chapter 14)
	40 Principles (Chapter 3)
	Trends of Technical Evolution (Chapter 4)
	Effects Database (Chapter 6)
	Smart Little People and Size–Time–Cost (Chapter 7)
	Resources and X-Factor (Chapters 5 and 6)
	9 Boxes (Chapter 8)
Rank solutions and implement	Ideality Plot (in this chapter!)

Depending on the kind of problem you're looking at (and how much time you have), it's also a good idea to add in a stage where you take your solutions and improve them. This thinking can be fast, and can create much better-developed solutions in a fraction of the time needed for the preceding steps.

Another reason you may want to develop your own problem-solving map is because you have other tools which you currently use with great success. However, you can also see that TRIZ has something additional to offer. Almost every workshop I've taught has had a Lean and/or Six Sigma expert in attendance, and many companies are bringing in TRIZ to work with other toolkits. If you have a good problem-solving system or process that works for you, I'm not suggesting you throw it away; you will have to work out how you fit TRIZ in with the tools you already use, though.

For Lean processes, I know many people use TRIZ for the 'Improve' stage of DMAIC (*DMAIC* is an improvement process and is an acronym that stands for Define, Measure, Analyse, Improve, Control). Generally, the totally unique solution tools in TRIZ are those elements that people want to add to their existing toolkit. All other problem-solving methods offer very logical and systematic approaches for understanding and defining problems, and for ranking and implementing solutions, but at the point of generating actual concepts or solutions they rely on brainstorming and other random methods. Only TRIZ fills that gap – but it has a lot to offer at the other stages of problem solving too. Many of the TRIZ thinking tools can provide great clarity of thought when added to traditional approaches; for example, consider completing a TRIZ Ideal Outcome to uncover all needs and stretch your thinking before doing more realistic and detailed requirements analysis.

Climbing the Problem-Solving Steps

This section runs through each step in the problem-solving process one by one. I prefer to think of these steps as akin to climbing a comfortable flight of stairs . . . or, if you're feeling energetic, scaling the heights of Kilimanjaro. Either way, focusing on one step at a time is the key to this.

Understanding the overall process

The overall process or route map for understanding and solving problems is shown in Figure 11-1.

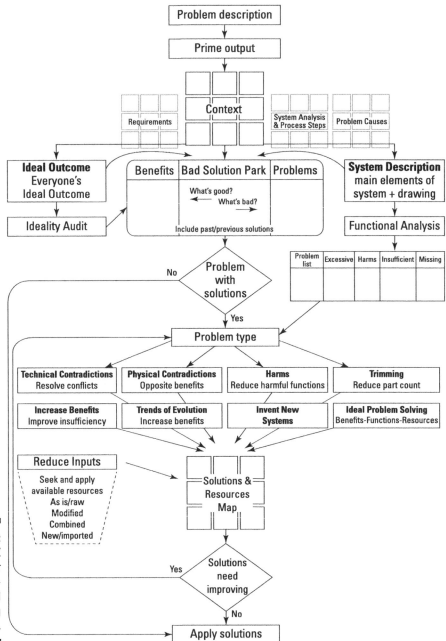

Figure 11-1:
The TRIZ process for understanding and solving problems.

Illustration by John Wiley & Sons Ltd.

Bad Solutions and ugly babies

Understanding some of the fundamental psychology about solutions is really helpful. You, like everyone else, love your own solutions – and that love for your own solutions is important.

A solution to a problem has to be worked on, developed and put into practice – and loving your solutions keeps you motivated throughout this process, which can be arduous. However, your love for your solutions can blind you to their downsides. Meanwhile, your colleagues may be able to see everything that's wrong with your solutions – and be happy to tell you!

TRIZ encourages you to accept your colleagues' opinion of your Bad Solutions, or 'ugly babies'. You can then acknowledge that they're not perfect and use TRIZ to turn them into better solutions (or pretty babies!).

The first stage of any problem-solving session is to create a *Bad Solution Park*: a place where all solutions can be captured and shared. They're called 'Bad Solutions' because you acknowledge that all top-of-the-head ideas have something really good about them but they can probably be improved. The term 'Park' is used because it's where you store your solutions until you need them, like a Lamborghini in a car park. See the nearby sidebar for more about Bad Solutions.

Some people dislike the term 'bad solution' and prefer to use terms such as 'initial solution'. I like Bad Solutions, however, because I feel it lowers the threshold for sharing them. Sometimes I have an idea that is really wacky and impractical – but I know it nevertheless has something good about it, and I want to share and discuss it with my colleagues. I can say to them, 'I've had a really terrible idea', and feel less embarrassed about sharing something that I know is half-formed and imperfect because I can then use all their cleverness and creativity to develop and improve my bad solution.

Sharpening the axe

Jumping straight to generating solutions without understanding the problem correctly is the biggest mistake you can make when problem solving. Unfortunately, this approach is particularly rife in management situations, hence the awful phrase, 'I want to hear solutions, not problems'. You love problems; you're a TRIZ problem solver!

If I'm facilitating a full day's problem-solving session, I typically spend the first half of the day understanding the problem – defining needs, understanding the current system and finding the gaps between the two to generate well-defined problems – but also collecting Bad Solutions from everyone because this process stimulates ideas (see the earlier 'Understanding the overall process' section for more about Bad Solutions). One client described these first few stages as a 'funnel' in which everyone's thinking converges, their attention is focused in the right places and ultimately the right kind of problems and their causes are identified.

I've noticed that teams of particularly sparky and creative people find the idea-generation stage of the TRIZ process the easiest because they're natural big-picture thinkers, who find it really easy to come up with lots of solutions. If left to their own devices, they generate hundreds of ideas; however, they can't identify which ideas to take forward because they haven't spent enough time understanding and scoping the problems. The first few stages in the process help them understand what problems they should tackle first so that they can then direct their creative energy into the most promising areas. These stages also give them criteria to help them work out which solutions should be taken forward.

Following definite steps is also useful for more methodical thinkers, because they understand why they're taking each one and what comes next. As a result, they trust the process and its logic and work more effectively. When I've attempted to 'short circuit' stages of the process, I generally meet with resistance. For example, my colleagues and I once ran a three-day facilitated TRIZ problem-solving workshop. After the first day, we came to understand and define the contradiction the team was facing and gave it to them to solve at the beginning of the second day. However, they refused to engage until we explained why we'd chosen that particular contradiction and 'showed our workings'.

Working together to understand and define the problem ensures that everyone understands the logic behind each step.

Make sure everyone is perfectly clear about which problem you're discussing. If you can reduce your problem description to a single sentence, that's great! 'Cut costs for Project X' or 'Create a better way of managing our projects' are good examples. If you're working in a team, give a short presentation or talk (ten minutes should be enough – no death by PowerPoint!) so that everyone is familiar with the area you're looking at. Have any necessary detail to hand (pictures are always useful; prototypes or physical examples are also help-ful if they exist), but you don't need to go into huge amounts of detail at this stage. The process will pull out all necessary detail as and when it's needed.

Prime Output is the main thing your system exists to deliver. For example, a restaurant's Prime Output may be 'the provision of delicious food and drink'. Although your Prime Output may seem obvious, it's important that you don't forget the main reason your system exists when you get into the nitty-gritty of problem solving.

Knowing your needs and scoping the situation

You can't define and solve your problem correctly until you've uncovered all needs. Do this stage very early in the process, before getting into the weeds and detail of your proposed or existing system. Some other problem-solving approaches suggest you define the problem before defining the Ideal, but I advise against that because it can result in psychological inertia; that is, you begin with the assumption that how things are currently done will form the basis for how things ought to be done in the future, which encourages incremental changes rather than innovative step-changes.

The TRIZ process will encourage both, but it is important to start with the big picture and then come down into the detail. For this reason, scoping out your problem first by creating a 9 Boxes context map (see Chapter 8 for more on the 9 Boxes) is often useful. The 9 Boxes context map ensures that you're tackling your problem at the right level. Things may be happening outside your original area of focus, which may actually be the right place to turn your attention; conversely, a detail may exist that, if tackled, will solve all your problems. Your 9 Boxes context map helps you chart everything that's happening and ensures that you tackle the problem at the right level.

If you're facilitating a problem-solving session, the 9 Boxes context map is also very useful for helping you understand the problem more broadly and the impact of different levels of scale, which not only ensures that you choose the right level to tackle your problem, but also that you have a record of why you chose it. One of my clients used the 9 Boxes to explain why his team chose to focus on a certain class of solutions; another used this tool to open up a conversation with senior management, as he wasn't sure at which level his team should be tackling the solution and wanted guidance.

Defining your problem correctly

TRIZ Function Analysis is the most important tool for defining your problem correctly. A client told me that he was having problems implementing the whole TRIZ process with his team. They'd mapped out the context of the issue, defined their Ideal, generated some solutions and looked at solving the contradictions. At this point, they'd diverged wildly and couldn't work out in which direction to go next. I advised the client to complete a Function Analysis to identify which parts of the problem were the most challenging. He could then see at a glance which parts of his system contained a lot of harms and insufficiencies, and could focus his attention on those parts of the problem. This approach allowed him to generate solutions in the most useful part of the problem, and also to converge the ideas of his team.

If you shape your solution generation around a Function Analysis, you can then see how to converge these ideas into a new, improved system; for which you can then draw up a new Function Analysis (and improve it again!).

This isn't to say that the fundamental problem in your system isn't a contradiction: sometimes people tackle problems with TRIZ because they've identified a contradiction and know TRIZ can help. In fact, this is the classic TRIZ approach; however, the reality of messy, real-world problems is that sometimes a number of interconnected problems are causing the issue, rather than one big contradiction. Function Analysis helps you see the wood for the trees: not only identifying one big contradiction but also pulling out all the other contradictions as well (the real problem may lie somewhere other than you originally thought), as well as any harms and insufficiencies (see Chapter 12 for more on how to deal with these).

The Function Analysis diagram or map (see Chapter 12) is also very good at communicating issues at a glance. At a recent problem-solving session, one team drew up quite a thorough Function Map of a road junction. While it was in some respects a very complicated diagram, what it showed at a glance was where two major problems lay. One problem was that various environmental elements were having a bad effect on the driver; the other was that insufficient maintenance could generate a large number of both direct and knock-on negative impacts.

Generating Solutions

When you have a well-defined problem, you can set about generating solutions. This is the most enjoyable part of the process and, uniquely in TRIZ, remains systematic.

Expecting the unexpected

When faced with a problem, you probably like to generate solutions. And you don't come up with solutions only when you're allowed to do so; you have ideas throughout the whole process, from first describing the problem and capturing your needs to understanding and defining the problem. You don't have to wait until the 'generate solution' part of the process to generate solutions; your brain will have come up with lots of ideas already by this point!

The problem-solving process isn't like a meal, where you have to eat all of your meat and finish your greens before you're allowed pudding (the bit we all enjoy most!). You can allow your ideas to flow at each stage of the process.

But – and this is a big but – you need to write them down and let them go. Don't hold on to them. Writing them down is important because they all have something good in them, but you want to move beyond these initial ideas and generate the right solutions to your problem. This requires a full understanding of the problem, and all your energy and attention needs to be focused on the stages that develop your understanding.

Writing down your ideas allows you to let go of them, and to come back to them later at the 'generate solutions' part of the process.

Also bear in mind that when you have ideas, they may take you off in unexpected directions. Problem solving is a bit like following a path through a forest – every now and again you spot an attractive glimmer among the trees. I encourage you to go and investigate that glimmer and to see what ideas you generate at that point, but, when you've played around with it for a bit, return to the path – the process. Never let anything get in the way of your free thinking, but come back to the process as soon as you've explored it. If you're continually going off in very new directions, that's another matter. Here, I suggest that you park those lines of enquiry and come back to them after you've completed the paths TRIZ suggested. You may well find that TRIZ has suggested all those ideas by the time you're at the end of the problem-solving process, together with other, equally useful solutions.

If you think of a solution, but can't see how it relates to the TRIZ process, it doesn't matter. Don't waste time trying to 'reverse engineer' your solutions into the process: the process has shaped your thinking and you've thought of a solution. That's great! The reason for doing TRIZ is to solve problems, not be a TRIZ wizard. Although that is also fun.

Applying the TRIZ solution tools

For many people, this is the fun part of the process. When applying the TRIZ solution tools, you're thinking freely, being creative and generating as many solutions as possible. You need to apply the principles of brainstorming (focus on quantity not quality, withhold criticism, welcome unusual ideas, combine and improve ideas) at this point so that everyone is willing to contribute. Thinking freely means that your creativity is being stimulated in the most useful places!

If finding solutions is like digging for buried treasure, the solution tools suggest where you should first apply your spade. The solution tools provide a catalyst for your creativity, and when you're trying to think of solutions, you're not doing so randomly; rather, you're starting in those places which have been statistically proven in the past to provide the most innovative solutions. This is why defining your problem correctly is so important: how you've defined your contradiction (Chapter 3) or the direction in which you've drawn your arrow in the Function Analysis (Chapter 12) will influence the kind of solution that's suggested.

Detailed explanations of how to apply the TRIZ solution tools can be found in the relevant chapters shown in Table 11-2.

Table 11-2	TRIZ Solution Tools by Chapter
TRIZ Solution Tool	*Chapter Number*
40 Principles for solving contradictions	3
Trends of Technical Evolution for future products	4
Resources	5
Database of Scientific Effects for 'how to?' questions	6
Creativity Tools	7
Standard Solutions for harms and insufficiencies	13
Trimming – for simplicity and cost reduction	14

Ranking and Developing Solutions

Generating solutions, as discussed in the preceding section, is only half the battle. The next steps towards implementation are equally important, and TRIZ can help. This section looks at *how* to implement those all-important solutions.

Developing solutions further

You've had some genius solutions. What do you do next? First, consider whether your solutions are perfect. If so, great, you can go home. However, I suggest that no matter how genius and apparently perfect your solutions are, they can probably be improved.

I recommend taking your solutions and running them through the process again. Time is relatively cheap, and because you're already thinking in a TRIZzy way, you should be able to move fast. Repeating the process need take only another hour, and that will definitely be time worth taking, particularly for important problems. If you come up with a more elegant, clever solution, you may find that it's quicker and easier to put into practice. You can also use this time to anticipate problems that may face your solution in the future – design or manufacturing problems, for example – and solve them right at the start of the process. Putting in time at this stage reaps disproportionately high rewards.

Ranking solutions by Ideality

If you have a difficult problem, being presented with 350 possible ways to resolve it isn't very useful! What you really want are three or four promising solutions to investigate further, and a method for ranking and sorting solutions is thus essential.

The TRIZzy way to rank solutions is according to their *Ideality*: what benefits each solution offers, and what costs and harms are associated with it:

- ✔ **Benefits:** You rank all solutions in relation to the benefits already captured in your Ideal Outcome.

- ✔ **Costs:** You consider not only how much money each solution will cost to create but also the time required to design, develop and test it. Obviously, the costs will be different for each solution.

- ✔ **Harms:** You take into account any undesirable outputs from the solution, including risks. These too will vary by solution.

You need to rank all solutions in this order: benefits, then costs and harms. Doing so is important, because you often reject solutions based on their problems although they nevertheless contain something really good – and your competitors do the same.

However, if you take those solutions most closely matched to your needs (they give you what you really want) but which may have big problems, TRIZ can help you solve any problems with the solutions and you may end up with very innovative and inventive solutions that competitors may miss.

Don't prematurely reject solutions because you think they're too hard to achieve. Someone else may know how to help, and that person might be in your organisation.

Using an Ideality Plot to rank solutions not only serves to converge your ideas, thus resulting in a smaller number to investigate, but also provides you with suggestions regarding what to do with the other solutions, as shown in Figure 11-2. You do this by:

- ✔ **Implementing:** Solutions with high benefits and few downsides (low costs and harms) can be put into practice immediately.

- ✔ **Solving problems:** These are the most exciting solutions; they have big benefits, but also big problems. These need to be worked on further by solving the contradictions (Chapter 3) or applying the Standard Solutions (Chapter 13) to deal with the harms. These solutions can form the basis of longer-term solutions. Bear in mind that the problems may be easier to resolve than you think and investigating one or two solutions from this category is always worthwhile.

✔ **Improving:** These are solutions that are easy to apply but have low benefits. They can be developed further using the Trends (Chapter 4), improving insufficiencies using the Standard Solutions (Chapter 13) or by combining them to create stronger solutions

✔ **Parking for now:** This is the polite term for solutions with low benefits and big problems. You probably won't need these: you will have a lot of better solutions you can use. If you really want to investigate them, you should apply both the strategies for solving problems and improving described above.

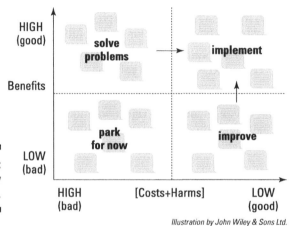

Figure 11-2:
The Ideality
Plot.

Illustration by John Wiley & Sons Ltd.

Converging by concept

If you still have a large number of solutions at this point, converging them according to their underlying concept can be useful. This helps you see not only the best broad directions forward, but you also have a number of different ways of moving in those directions.

Any problems with individual ideas can then be solved as they occur. You're not stuck in the details of any specific solution, but focused on how to achieve the outcomes you want.

Solving Difficult Problems Effectively in a Team

TRIZ works well for individuals but better for teams. Very rarely in life do people experience problems for which they have to find solutions on their own. Creative teamwork becomes easy when people use a systematic process together, but in order for that process to work, they also have to be aware of some of the psychological aspects of teams.

Gaining team understanding and consensus

One of the most important elements of understanding and solving problems is working as a team. Two heads are better than one, and five are even better. Seeking your colleagues' help and advice – within the structured process of TRIZ – is a good idea because they'll bring fresh thinking and new ideas. However, when you're working on a team issue, it's essential that the whole team is involved, especially those people responsible for implementing solutions.

Not involving the people who'll be implementing the results of a problem-solving session is a huge mistake. No one appreciates having solutions handed to them on a plate. If they're involved in the development of these solutions, they'll feel a sense of ownership of them and will be more enthusiastic about putting them into practice (and more likely to do so!). More importantly, they'll also have valid contributions to make. Involving those who have to implement the solutions may seem common sense but I've encountered teams who deliberately exclude these people because they 'see problems everywhere'. That's exactly the kind of thinking you need in a TRIZ session! These people can point out all the potential problems as you're working on the issue and you can then address them upfront.

Converging and diverging

As you work through the TRIZ process, you'll experience moments of divergence and convergence. You need both.

Diverging into ever-more solutions isn't helpful. You need to focus on a manageable number of solutions to explore and test. Understanding the cyclical nature of the problem-solving process enables you to understand the outcomes for each stage; that is, you needn't panic that ideas are diverging because the next stage in the process will converge them again.

You initially diverge when you explore all needs, then converge when you distil them into a defined Ideal Outcome. Divergent tasks include problem exploration and understanding of the context; convergence involves detailed understanding and definition of the problem. When you have a well-defined problem, diverge to think of many solutions; when you have many solutions, you evaluate them and select the top few to take forward.

If you have many solutions, you can start to converge them by grouping the fundamental concepts behind them; you can then evaluate these concepts according to their Ideality and create a shortlist of the most promising to take forward and a clear plan of what to do next.

Implementation requires diverging to explore, investigate, improve and develop. As problems arise (as they always do), the solutions are whittled down, problems solved and a better solution finally put into practice.

Tasks can be convergent or divergent, and require different thinking approaches. Most of your working life shows you how to be good at convergent tasks; those requiring planning, making things work, finding problems, identifying constraints, selecting and developing shortlists and so on. Divergent thinking involves generating lots of new ideas and making new connections. Creativity and innovation training therefore usually focuses on divergent thinking, because if you're successful at work, you're probably already good at convergent tasks. Real innovation, however, requires both kinds of thinking.

Thinking fast and thinking slow

Daniel Kahneman's book *Thinking, Fast and Slow* (Farrar, Straus and Giroux, 2011) contains a fantastic description of two fundamental types of thinking that you use when solving problems.

Broadly speaking, the two ways of thinking may be defined as follows:

- ✔ *Thinking fast* means having a flash of inspiration or making an intuitive leap. It tends not to involve conscious processing; you suddenly leap to a solution and you're not sure exactly how or why you ended up there. Never suppress this type of thinking! Allow it to happen but don't rely on it; because it's largely unconscious, it's difficult to predict – inspiration may or may not strike. Kahneman also discusses the fact that fast thinking can trip you up as a result of bias; TRIZ describes this experience as a kind of psychological inertia.

- ✔ *Thinking slow* means systematically stepping through a problem situation to break your psychological inertia, understand the real problem and generate solutions. Thinking slow means you'll cover the whole problem space – and the whole solution space.

You need both kinds of thinking. The more experience and expertise you have in a particular field, the more likely you are to be able to jump very quickly to solutions based on previous success. However, this approach can be very bad for innovation because it tends to take you to solutions you've thought of before. Your expertise is thus both good and bad. Fortunately, following an explicit problem-solving process ensures that you cover all elements of the problem, and the TRIZ solution tools point you to places that you may not have considered on your own. You can then use your intuitive, fast thinking to generate more solutions in the places where you're most likely to find innovation. Ultimately, you get the best out of both kinds of thinking.

Some people are much happier to think intuitively while others like to be more systematic. As teams need both kinds of thinking, it's important to appreciate and allow both – and to encourage people to work and participate in the way they feel most comfortable. This is where working in teams of three to four people can be so valuable. When the team is dealing with a task that requires rigorous, detailed and systematic thinking, the person who feels most comfortable working in this way can lead and take responsibility for the successful completion of that task. Those who think more intuitively can participate and also helpfully 'play' around the problem, suggesting new approaches and generating solutions. When brainstorming new ideas, the more intuitive thinkers can lead and stimulate fresh thinking.

Putting solutions into practice

If you follow the steps outlined in this chapter, putting your solution into practice should be more straightforward than would otherwise be the case. You'll understand the problem and ensure that you find solutions to the right problem. You'll generate many innovative solutions, then rank and select them. The ideas that you generate should be sufficiently exciting and practical that the first steps are obvious.

By stepping through the problem with the right people in the room, you'll secure the engagement of those responsible for implementing the solutions; they'll also be enthusiastic about putting these ideas into practice. You'll also create a useful resource to draw on when the inevitable problems emerge; these people will be happy to help you deal with these problems and develop the solutions.

A great deal of psychological research has investigated the personality traits required for innovation. Alongside motivation, persistence and social skills is the ability to enable others to understand why your solutions are good and should be put into practice. You can use the TRIZ process to engage all the relevant people and ensure your wonderful and innovative solutions will be taken forward and thrive.

If you have difficulty implementing your solution, possibly you haven't considered all the relevant stakeholders. To develop and implement any solution, multiple stakeholders need to engage with it. They also need to be able to move away from the old way of doing things. Fantastic solutions make life easier for everyone, and the idea will sail through implementation.

Some solutions, however, may make life difficult for someone far removed from you, and if the success of your solution requires someone to do more work on a consistent basis, he'll resist. Focusing only on the people near you or your immediate customer is common. But if your solution hits a stumbling block, solving the problems of your customer or supplier (using TRIZ!) is in your best interests because doing so will allow your solution to thrive.

Solving other people's problems

The great success of Amazon's Kindle compared to the Sony Reader can be attributed to the fact that Amazon solved the problems of both its customers and suppliers. The Sony Reader was the first e-book to be released and was a beautiful invention, with a very easy-to-read, bright screen. Amazon's Kindle was inferior technology: it was bigger, heavier and the screen wasn't as good. However, two problems faced this new industry: publishers were concerned that e-books could be easily copied; and customers wondered how they'd access and pay for them. Amazon solved these problems by creating a rigorous closed system to prevent piracy, which kept the publishers happy, and providing a one-stop shop for customers, which made downloading and paying for e-books very simple. As a result, the Kindle quickly became the market leader. Amazon displayed some very TRIZzy thinking, looking at problems more broadly and understanding what their supply chain and their customers' concerns might be -- and tackling them in advance. Many successful businesses display this clever thinking, both identifying and solving other people's problems for them.

Chapter 12

Getting to Grips with Your Problems with Function Analysis

In This Chapter

▶ Turning messy situations into a well-defined problem list

▶ Understanding problems and the best routes forward for solving them

▶ Describing and communicating issues so anyone can understand them

*O*ne of the joys of TRIZ is that it takes complex problems and breaks them down into a series of simple and easy-to-understand relationships. Function Analysis maps these messy, complex issues in a diagram that can be understood at a glance, and difficult problem solving is broken into small, easy-to-digest chunks.

This chapter looks at how you can use Function Analysis to your best advantage.

Making Complex Problems Simple

The way *Function Analysis* works is to take complicated situations and break them down into a series of binary relationships. You build a diagram or *map* of these interactions, which describes how your system works taken as a whole at one moment in time – this enables you to tackle the big, messy system one simple interaction at a time.

Function Analysis gives you focus: you concentrate on the essential problems without being distracted by irrelevant detail or psychological inertia. Function Analysis helps you not only see the big picture of how your system works but also to zoom into the essential problems, define them correctly and then find new solutions by applying the TRIZ Standard Solutions.

Let's go through a complex real-life example and see how completing a Function Analysis can help you understand the situation. The Function Analysis shown in Figure 12-1 describes the situation in many hospital intensive care units (ICUs). Patients in intensive care are very sick, and often on ventilators to help them breathe. Because they're very sick, they can get much worse very quickly and thus need to be monitored on a continuous basis. For that reason, they're attached to machines that monitor their heart rate, blood pressure and breathing. If any of these changes suddenly, an alarm goes off, notifying the doctors and nurses that the patient needs attention. However, when an alarm sounds, it also disrupts the patients who are asleep, which is a problem because sleep is essential for their recovery. Doctors and nurses also talk, which is good in terms of sharing essential patient information and fostering a sense of teamwork, but their conversations disturb patients. Good teamwork has a significant positive impact on patient outcomes, however, so staff interaction needs to be managed not discouraged.

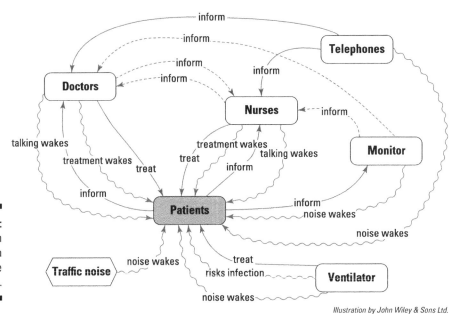

Figure 12-1:
Function
Map of an
intensive
care unit.

Illustration by John Wiley & Sons Ltd.

What this Function Map tells you is that ICUs are very noisy places, which is bad for patients. However, the noise occurs as a side effect of useful actions being performed – and the information provided by the machines and staff interaction must be captured.

Looking at a diagram like this helps you understand what's really happening. Even if you're not an intensive care specialist, a Function Analysis diagram allows you to understand what problems are occurring and why; it's an essential communication tool. When you've completed a Function Analysis, you can then look to Trimming (Chapter 14) and the other Standard Solutions (Chapter 13) for suggestions on how to deal with the problems.

Understanding the building blocks of Function Analysis

You should use the simplest language possible when defining your actions; a four-letter verb of one syllable is the ideal. The verb must differ from the subject or object: 'printer prints paper' is tautological and will result in psychological inertia – 'printer marks paper' is better functional language.

Function Analysis consists of a number of Subject–Action–Object relationships, as shown in Figure 12-2. A function delivers an outcome we want and is achieved when a subject (or thing) performs an action on an object (another thing): in this case, the doctor is treating the patient. The subject is the thing doing the action (the doctor); the object is the thing that's changed (the patient); and the action is what performs the change (treatment). The function is the combination of the action and the object together: in this case, 'treat patient'.

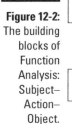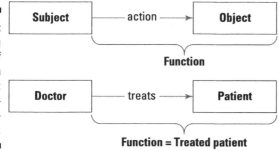

Figure 12-2: The building blocks of Function Analysis: Subject–Action–Object.

Illustration by John Wiley & Sons Ltd.

Getting this relationship the right way round is important, as it helps you understand how your system is actually working. It's also critical for the problem-solving stage, when you apply the Standard Solutions. If you were to say 'I love George Clooney', is that a correct functional description? While it's correct in normal language, it doesn't work for Function Analysis: George Clooney isn't changed by your love (unless you camp outside his house and send him letters every day). Rather, the correct description of the function might be 'George Clooney inspires love in you'. You must be the object of the action, as you're the thing that's changed.

The Function Analysis can be as big or small as required: I've seen a Function Analysis of a submarine, which was very comprehensive and very complicated. A good rule of thumb, however, is a Function Analysis with about 10–15 components that's easy to understand at a glance. A broad, high-level Function Map can show you where you need to zoom in: you may want to create an overall Function Analysis of your whole system, and then do more detailed analyses of specific troublesome components. This approach allows you to explore the detail – and keep it in its place.

Know the difference between Function Analysis and a Function Map:

✔ *Function Analysis* is the process of understanding your system by focusing on its functions

✔ A *Function Map* is the diagram you draw, which shows all the components and how they interact and all the functions, both good and bad

A Function Analysis can help you understand how your system really works, which is useful for large and complicated systems in which individuals may be working in just one small part. Understanding how the bit of the system they specialise in interacts with other parts can be incredibly useful and generate great insight – even in teams that have been working with their systems for years.

Defining interactions: The good, the bad and the ugly

Four types of interaction exist in a Function Analysis, as shown in Figure 12-3.

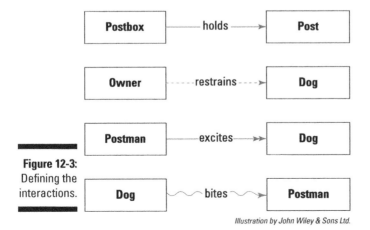

Figure 12-3: Defining the interactions.

Illustration by John Wiley & Sons Ltd.

These interactions are:

- ✔ Useful actions, shown by a solid line
- ✔ Useful actions that are insufficient – that is, give you something you want but not enough of it – shown as dotted lines
- ✔ Useful actions that are excessive, shown as double lines or lines with double arrow heads
- ✔ Harmful actions, shown as wavy lines

When working in colour, it's helpful to make useful lines green and harmful lines red because then problems stand out much more clearly.

It's worth noting that, of the four types of interaction revealed, three are problems. That's great because every identified problem is an opportunity for improvement. When you've created a Function Analysis, you can write a problem list of every interaction, revealing insufficient, excessive or harmful functions, and then work through it solving problems one by one.

The only other kind of action that can sometimes be useful to capture is a completely missing useful action. Although it's a bit theoretical when looking at real systems with real problems, it's very useful when describing how a new system may look, because other things you'd like may occur to you as you work through the Function Analysis. Missing actions can be drawn and treated as insufficient actions; so insufficient they don't yet exist! You may find identifying them using slightly different dashes or in blue, rather than green, if you're working in colour, helpful.

TRIZ Function Analysis versus other methods

TRIZ Function Analysis is a simpler form of the TRIZ Substance Field Analysis. Both connect identified problems to the TRIZ Standard Solutions. The linking of TRIZ Function Analysis to the TRIZ Standard Solutions has evolved in different ways within the TRIZ community. This uses the Oxford TRIZ Standard Solutions which were rearranged according to the kind of problem they help you solve.

TRIZ Function Analysis is more rigorous and often a different logic than other methods which also use the name Function Analysis (such as in Value Engineering). Other forms of Function Analysis map the interactions between different components, to describe how a system works as if in a perfect world. TRIZ Function Analysis also allows you to map all the places where your system isn't giving you everything that you want – *all* the problems.

Charting all parts of your system and beyond

When drawing a Function Map, include a number of different types of components, as shown in Figure 12-4 and described below.

- ✔ **System components:** Components over which you have control, depicted as white rectangular or oval shapes.

- ✔ **Environmental components:** Components that interact with your system over which you have no control or can't change, shown as hexagons.

Figure 12-4: Different types of component addressed in a Function Map.

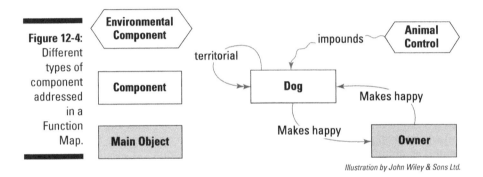

Illustration by John Wiley & Sons Ltd.

You need to include environmental components because they affect your system. It may actually be possible to find ways to deal with the problems they create without changing the environmental components at all (the Standard Solutions, in Chapter 13, show you how).

Making sure you capture the interactions of the environmental components with your system is really important, as they'll show you opportunities to improve the parts of the system you can control. Environmental components may not only be simply part of the environment itself (for example, gravity, sea water), but also other elements that you don't control, such as your user. It's important for product designers, for example, to consider all the ways in which a user could misuse or damage a system and attempt to mitigate these in advance.

The last thing to bear in mind when completing a Function Analysis is to make sure the Prime Function of your system has been captured.

The *Prime Function* is what your system exists to deliver: its main purpose. You can write this as a Subject–Action–Object: the object is the 'main object', or target, of this function. The main object is what's intended to be changed by the system you're describing; it may be the user of a system or, in the case of a technical system, what your system exists to change.

For the intensive care example shown in Figure 12-1, the Prime Function is 'treat patient'. In this case, two components are delivering the Prime Function: doctors and nurses. The main object is the patient: what you want to be changed by the system, and the reason for the existence of the system.

How you define the main object depends on how you define the level of the system. For example, if you take an insulin injection device as your system, the Prime Function is 'move drug'; if you take the treatment of diabetes as your system, the Prime Function is 'treat patient'. Function Analysis at both these levels is interesting and useful. What you choose depends on what you're hoping to achieve in your problem solving. Thus, if you're a medical device designer wanting to create a new insulin injection system, you'll choose the former; if you're a healthcare provider looking to improve treatment overall, the latter. The process of identifying the Prime Function also helps scope the level at which you want to examine the problem, which is an important decision. Carrying out more than one Function Analysis at different levels may be worthwhile, as different kinds of solution will be suggested.

When in doubt, start high level. It's easy to add more detail later but hard to remove detail once it's in there. What's important to remember is that any Function Analysis should be focused at just one level.

You can look at a Function Analysis of the action of insulin itself, or up a level at the injection system, or up another level at the diabetes treatment. However, you shouldn't mix up these levels within a single Function Analysis. Starting high level means your Function Analysis can indicate where the biggest problems are, which can help you choose where to focus a more detailed Function Analysis (for example, perhaps an insulin injection device works pretty well, but problems exist with blood glucose measurement, patient compliance or communication with healthcare professionals in the whole treatment system of diabetes).

In Figure 12-4 I suggest 'dog makes owner happy' as the Prime Function of the system (some dog-lovers may feel it's the other way around!). Keep in mind the main reason your system exists: to deliver the Prime Function.

Another kind of action is also worth mentioning: inherent features of your system, as shown by the recursive action of 'territorial' on the dog in Figure 12-4. Sometimes components have inherent features that you need to consider; and these can be useful (for example, your dog could be inherently affectionate), harmful (it could be aggressive and sometimes bite you), insufficient or excessive (as in this case). Capturing these inherent features is worthwhile because they carry useful information about individual components, and tell you which parts of your system are inherently good, insufficient or bad (you can include harms such as dangerous and heavy, for example).

Building a Function Analysis Diagram

Function Analysis is the most powerful tool I know for defining and therefore truly understanding problems. Through mapping out how a system works – and what's going on – teams not only understand what's really happening but also are confident that the complicated reality is captured in a diagram.

Seeing problems in a more pictorial format helps you restructure your view of the problem and breaks your psychological inertia. Function Analysis distils your problems into a series of simple interactions, but that doesn't mean the process is always easy. It often takes significant brain power to understand exactly what's going on in your system, and to draw the system boundaries. This thinking is done in a structured way, however, and the process of working through these issues gives great clarity of thought and a better understanding of your problems.

Completing a Function Analysis

In essence, completing a Function Analysis is as simple as following these six steps (the reality, of course, requires a little more input!):

1. **List all your components.**
2. **Create a list of all functions each component delivers and a table of all interactions.**
3. **Draw a Function Map.**
4. **List all problems.**
5. **Prioritise problems and solve: trim, apply Standard Solutions and solve contradictions.**
6. **Take the new improved system and repeat the process.**

The following sections discuss each step in more detail.

Listing your components and their interactions

The first stage in building a Function Analysis is to list all your components – system and environmental – for example, doctors, nurses, patients, traffic noise.

Creating a list of all the functions each component performs before drawing the Function Map can be useful, because it ensures that every function is captured. You start by picking a component, for example 'doctor' in this example, and listing all the actions a doctor performs and on what components of your system.

Another useful step is to create a table of all interactions between the components of your system. The first stage is to identify whether each component does or could physically touch or directly interact with each other component in any way. This table doesn't tell you what the interactions actually are, but it does form a useful memory aid when you're completing the Function Map to ensure that each interaction has been captured. In the ICU example, phones are likely to interact with nurses, doctors and patients; they won't interact with monitors, ventilators or traffic noise.

Drawing a Function Map

At this stage you're ready to draw the map or diagram of all the functions.

Put your components on sticky notes first, so you can move them around. Place components next to each other that are physically close in reality because they're likely to have more interactions.

If you've chosen to jump straight to drawing the Function Map without creating a function list, you can still double-check you've captured everything by picking a component and pointing at all other components one by one and asking whether further interactions exist between them. This is a quicker approach and what I usually do in live problem-solving sessions when time's short. People often find the action of drawing the Function Map both satisfying and comforting and want to get to it quickly. It starts to draw out problems and represent them clearly, and they'd rather start with a 'quick and dirty' Function Analysis that they can then improve and refine.

Listing, prioritising and solving problems

You then create a problem list: every insufficient or excessive action and every harmful action can be written down in a list for you to tackle one by one.

This is the really fun bit! You take each problem and apply the relevant problem-solving tools. First, trim your system (see Chapter 14), then apply the Standard Solutions to any insufficient or harmful actions (Chapter 13). If useful and harmful actions appear together, you have a contradiction (see Chapter 3 for what to do about contradictions).

Prioritising is worthwhile; often it's very obvious where to start. You may find that some areas are crowded with problems but, if not, start with the problems that are nearest to your Prime Function (what your system is supposed to deliver) and work out from there.

You'll then generate a range of solutions, which you should bring together into a new, improved system.

And then, the final step (kind of) . . . do the same again! Identify all the components, draw another Function Map and improve. This stage will take your ideas much further relatively quickly.

Uncovering Conflicts: Putting Contradictions in Context

Solving contradictions is one of the most widely known of the TRIZ tools, and the idea that you can solve contradictions is one of the parts of TRIZ thinking that's so unique and powerful that everyone gets excited about it. When you have a really big underlying contradiction, it must be solved if you're going to improve your system. Sometimes, however, contradictions are only a part of the problem. An example of a contradiction described in the Function Analysis in Figure 12-1 is that a doctor treating a patient is a useful action but treatment also wakens the patient, as shown in Figure 12-5.

Figure 12-5:
A contradiction described in Function Analysis.

Doctors — treat → Patients
treatment wakes

Illustration by John Wiley & Sons Ltd.

You could describe this contradiction as simultaneously wanting the doctor to treat the patient and not treat the patient (for example, you could solve this Physical Contradiction by separating in Time, and just treat the patient at certain times of day). This is a very difficult contradiction to tackle because 'doctor treats patient' is the Prime Function for your system, and the constraints you are operating within may make it impossible to completely resolve the contradiction in reality. Your Prime Function in the case of 'treat patient' will trump all other considerations. Also bear in mind that many other things in your system are also waking the patient (including the nurses providing treatment, the conversation between the doctors and nurses and the noise from the instruments). Also, 'treat patient' as a function is high level and contains many different actions. It could be useful to zoom in and consider all the different actions performed by the doctors and nurses to treat the patient; for example, they may be injecting the patients with drugs, moving them to wash them or rolling them to prevent pressure sores, all of which disturb the patient in different ways. Considering all of these separately may provide more detail about how you could mitigate the harms individually. Looking at the harms enables you to leave your Prime Function intact and allows you to deal with *all* the problems in your system – not just the main one.

Completing a Function Analysis thus helps you find all problems: contradictions *and* harmful actions, insufficient useful actions *and* excessive useful actions. You can then not only solve contradictions but find other ways and places to improve your system, enabling more comprehensive problem solving. It also helps you find the best place to solve problems, if you have a contradiction associated with a particular component, removing that component may be easier than trying to solve the contradiction.

Function Analysis also helps you see how things interact and connect. Often, problems are a result of a number of things happening, and you need to tackle them all if you're going to create a much improved and more robust system. Function Analysis ensures you capture all problems so that you can generate as many good solutions as possible.

Understanding How Everything Fits Together

One of the unexpected outputs of Function Analysis is that everyone sees how systems fit together. Problem-solving sessions often end with participants saying the most important part of the process was completing the Function Map, as it helped them understand what every element of their system did, and how things fitted together (often the most important part was the understanding generated – not the map itself).

When you truly understand how a system works, you're able to make intelligent decisions about how and where to change it. You're also able to see the interaction between the useful things that are going on alongside your problems. Sometimes you can become so focused on particular problems that you're willing to do anything to get rid of them, and can risk throwing the baby out with the bathwater. A Function Analysis helps you see how everything is interconnected so you make sensible changes and ensure that you make things better, without anything getting worse.

By understanding how your system operates, and understanding all the good and bad elements, you can take a step back and see what's really at the heart of the issue. People have a tendency to over-develop what they know and understand and find interesting, and also to focus most of their attention on the parts of the system most directly under their control. However, the real heart of the problem and the biggest causes could still be something you can make positive changes to prevent or mitigate, if only you know that's where you have to look.

The first time I carried out a Function Analysis with a group of engineers who designed injection devices for insulin, they commented that what was most interesting was that the majority of problems they identified concerned the needle (which is dirty after use, dangerous to the patient and others, hurts the patient during injection and so on). They said that, while they always knew this was the case, it was rarely the focus of their attention, as they spent most of their time designing the things that went around the needle; as mechanical engineers they were most interested in (and indeed responsible for) the technical functioning of the device. However, they realised it was worth reminding themselves that the needle is a big source of problems, and many opportunities for them to mitigate these problems in the device design existed. The Function Analysis, which identified a large number of harmful actions associated with the needle, reminded them of this fact.

Sometimes when you step back after completing a Function Map, you have a clear visual picture of the problems. Often, lots of harms and insufficiencies will be focused in certain areas and the source of the problems is perfectly clear – and in a manner that's easy for everyone to understand.

Using Function Analysis

Function Analysis is essential when you face difficult problems. Because it helps you not only understand but also define problems, it is particularly useful when you are struggling to see where to start with an issue. It is useful, therefore, not only for problems that start out relatively well-defined (for example, something's started leaking: we need to find the leak and fix it) but also for complex issues where gaining understanding is almost as important as finding solutions (such as potential causes of global warming) and very challenging problems that will require multiple lines of attack (for example, reduce costs by 50 per cent).

Modelling difficult problems

Most problems that you tackle with TRIZ are difficult and start out messy. You often read about people solving problems by starting from a general description of the area and then leaping to a well-defined problem and working from there. The magical leap to that well-defined problem isn't actually described, however. Looking back at problems gives you 20:20 vision: it's perfectly clear in retrospect why you tackled one particular part of an issue, but looking forward things can seem less certain.

When you're real-life problem solving, it's okay to assume you're starting with a great big mess. Stepping through the problem-solving process will help give you clarity of thought, and the most important stage in the process is the Function Analysis. Often, when working on a problem with colleagues, people will disagree about its source. In reality, a number of different causes of problems will probably exist and Function Analysis will capture them all.

A functional description of every part of your system when put together can help you really understand how your system works – with great clarity.

Very, very rarely does just one big cause of the problem exist; more usually, a complex series of interactions combine to create a number of problems at different levels of severity. Capturing all these interactions in Function Analysis, and all the problems, is useful because then you've taken a difficult problem and broken it down into a number of smaller, well-defined problems that you can tackle one at a time.

Part of that clarity of thought derives from the effort you take to define each action correctly. It's very easy to assume you know how everything works, but having the discipline to step through each part of your system and describe each interaction without using any technical jargon but just very simple terms can be surprisingly hard.

For example, taking the ICU example, what does a heart monitor do? If you're using normal language, you'd say a heart monitor measures the heart rate. This, however, isn't a functional description. Remember the rule: a subject has an action on an object. The object is the thing that's changed – is the heart rate changed by the monitor? No – the patient's heart rate informs the monitor. A more precise description is shown in Figure 12-6.

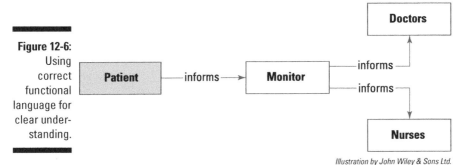

Illustration by John Wiley & Sons Ltd.

Figure 12-6:
Using correct functional language for clear understanding.

When you complete a Function Analysis, you're also taking very difficult problems, modelling their potential causes and understanding exactly why they're problematic. You can then tackle them one by one. Your difficult problems are turned into a series of smaller problems, which you can list and then tackle individually. These smaller problems – the binary interactions you've modelled in your Subject–Action–Object relationships – are very well-defined, much easier to tackle and much less intimidating. You're not trying to think of solutions to massive problems all at once (for example, improve patient recovery times). You just have to think of solutions to this one particular small problem – for example, stopping the noise created by doctors and nurses talking from bothering patients – and the TRIZ tools, particularly Trimming (Chapter 14) and Standard Solutions (Chapter 13) provide the routes to change things.

Gaining the confidence required to change things

When you've completed a Function Analysis and really understand your problem, you can start problem solving and generating new solutions. The first thing that generally happens when someone suggests changing something is someone else saying, 'you can't do that because . . .'. Changing or even removing parts of your system will probably have knock-on effects. People resist change because they're familiar with the old way of doing things, even if it has problems; doing something different may result in new, unexpected problems. What Function Analysis gives you is not only confidence that you really understand what's happening in your system but also the ability to predict the result of any changes you make and to remodel your new system in Function Analysis.

The Trimming Rules are a very powerful approach to problem solving, which suggest improving your system by removing rather than adding things (see Chapter 14 for details). This approach may seem counterintuitive, but what the Trimming Rules allow you to do is keep all the useful actions of components with problems but remove the components; you thus have all their benefits but with fewer costs (as you have less stuff in your system) and fewer harms (as any problems associated with the component will disappear because the component is no longer there). Your Ideality has increased, as shown in Figure 12-7.

Figure 12-7:
Improving
Ideality with
Function
Analysis.

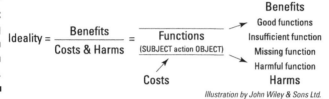

Illustration by John Wiley & Sons Ltd.

Function Analysis allows you to increase your Ideality by focusing on functions. Your costs – inputs – give you functions. These functions have two outputs: things you want (benefits) and things you don't want (harms).

Function Analysis helps you map all these functions and then work out how to improve your system by improving their functions. If you can remove components but keep their useful functions (which are giving you benefits), then your system is more ideal. When people throw up their hands in horror at the suggestion of removing something from the system, and list all the things a particular component is needed for, both its useful functions and the benefits those functions deliver, you can reassure them that Function Analysis has captured them. You can then follow the Trimming Rules to transfer those useful actions to other parts of your system. When you've done that, your team will look and see that, yes, actually, that component can be removed because everything useful it does has been kept, but what emerges is a simpler system. As a result of stepping through the process in a systematic and logical fashion, the whole team feels confident that the changes can be made.

If you're worried about how your system will function as a result of changes, you can draw another Function Map, plotting out and imagining how your new system will work, functionally. You can identify any problems, both old and new, and then apply the same problem-solving tools to get rid of those problems. What people can see when a new Function Map has been drawn up is how a new system can work (and, being realistic, all its potential problems), and they can feel confident in the solutions generated.

Achieving clarity in complex situations

One of the benefits of Function Analysis as a method is that it helps you see the wood for the trees. You start with a big mess but that big mess becomes a number of simpler, binary interactions which we can see at a glance on our Function Map.

What you're doing is taking a complex, messy situation and, in your Function Analysis, making it merely complicated. For a very big problem, your Function Analysis may still be a tangle of arrows between components, but you've removed some elements of uncertainty. What was messy and full of unknowns has become a diagram or map of all the problem areas in your system.

Those problems are defined as simple, Subject–Action–Object relationships. You can list each Subject–Action–Object relationship that's problematic (whether it's insufficient, excessive, missing or harmful) and then tackle each simple problem one by one. Your complex situation is now broken down into manageable, bite-sized problems that you can work through logically. Working on each individual problem is simple.

Clarity of thought is also achieved by thinking functionally. Focusing on functions strips out all the unnecessary detail of a real-world problem, and provides a way of thinking that's both more abstract and very precise. Clever problem solving takes place at the function level, which is one aspect of engineering thinking that the rest of the world can benefit from. Understanding what functions are required is the first step to seeing new possibilities, as there will be many ways of delivering the function that you want, beyond the ways in which we currently deliver them (see Chapter 5 for more on how to systematically search for new ways to deliver functions).

Describing your problem visually also has lots of benefits. It helps you break out of your normal way of looking at the situation and bring new thinking to the problem. Colour-coding harmful actions means you can see at a glance where your biggest problems lie. Some components may be in the middle of a tangle of red arrows, which is a sign that that middle point is a useful place to start your problem solving.

Communicating effectively

Function Analyses are very powerful for communicating problems. With only the briefest of introductions to the tool, it's possible for people with no experience or knowledge of TRIZ to understand what the Function Analysis is saying and where the problems lie. My company worked on one project that highlighted the need to create a new military regulatory system for the development and maintenance of military aircraft, which was communicated in a Function Analysis so simple even a politician was able to understand it at a glance!

If it's true that you only really understand something when you can explain it to someone without specialist knowledge, the discipline of completing a Function Analysis with no technical jargon or language ensures you'll be able to do so. Using very simple language when defining the actions needing to be taken also means you'll be able to engage people in developing solutions outside of your area of expertise. This will also result in the creation of diverse teams: a Function Analysis describes all the problems in simple enough terms so anyone can understand them, which means that everyone can get involved in generating solutions.

A completed Function Analysis not only describes the current situation but also outlines the need for change: you're highlighting all the good things that are happening *and* all the problems. Human beings are problem solvers: when they see a problem, they start to generate solutions.

When you've described and defined the problems, you've engaged your audience in finding solutions, and you can then move on to creating a better system by applying the solution tools (starting with Trimming, which is described in Chapter 14).

Chapter 13

Solving Problems using the TRIZ Standard Solutions

*T*he Standard Solutions are one of the problem-solving tools developed from the categorisation of solutions to existing problems and how those problems have been solved inventively: think of them as a kind of 'library' of answers that the world has generated in the past! The Standard Solutions form one of the broadest solution tools in TRIZ, encompassing many of the other tools based on patent analysis (the 40 Inventive Principles, the Trends of Technical Evolution and the Effects Database).

The *Standard Solutions* are simple TRIZ lists for dealing with problems: one of the tools developed from patent analysis (as described in Chapter 1) which equip you with known, standard ways of dealing with harms, improving insufficiencies and measuring or detecting things.

Defining a Subject–action–Object

In order to apply the Standard Solutions you have to define your problem as a single Subject–action–Object. When you have your Subject–action–Object (SaO) in order, you can then break it down into three components:

✔ **Subject:** A component which is delivering an action (for example, an oven)

✔ **Action:** A change being made or a 'doing word' (for example, heats)

✔ **Object:** A component which is changed by the action (for example, food)

You must get your Subject and Object the right way round, as described in the Trimming Rules (Chapter 14), because this makes a big difference to how you look for solutions. Most importantly, it also helps you understand clearly how your system works, what is providing functions and what is being changed.

The Standard Solutions suggest changing the Subject, the action and/or the Object. For this reason, your problem needs to be structured at least as a single SaO, if not as many linked SaOs (a Function Analysis, as described in Chapter 12). Without an SaO, the Standard Solutions suggest only approximate solution directions as you have nothing to which you can precisely and most effectively apply them.

An SaO can describe only four kinds of action:

- ✔ **Useful actions** which deliver something you want
- ✔ **Insufficient useful actions** which deliver something you want but not enough of it
- ✔ **Excessive useful actions** which deliver too much of something you want
- ✔ **Harmful actions** which deliver something you don't want

Only useful actions are without problems: the other three all point you towards improving the actions to get just what you want (no more, no less). You shouldn't be afraid to identify problems, in fact I would encourage you to seek out as many problems as possible, as each provides an opportunity for improvement. If you're not sure whether something is useful or insufficient, for example, err on the side of caution by assuming it's a problem and mark it as insufficient. This will provide you with a prompt to find ways to improve.

Some examples of modelling some real-life situations as Subject–action–Objects are shown in Figure 13-1.

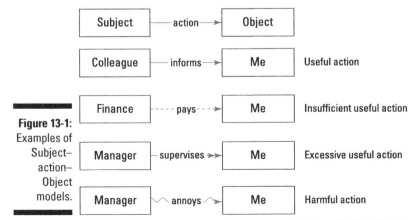

Figure 13-1: Examples of Subject–action–Object models.

Illustration by John Wiley & Sons Ltd.

When you've defined your SaO, you've described your problem in a simple model. You can then apply the Standard Solutions to deal with the problem and improve your system!

Categorising Problems

In total, there are 76 Standard Solutions. They offer a different angle to other tools, as they categorise problems into broad classes and provide a number of literally 'standard solutions' that have been used repeatedly to solve these kinds of problem.

The traditional Russian approach has five classes of solutions organised partly by problem type and partly by solution type (see sidebar), but in Oxford TRIZ these have been rearranged into three classes of solution according to the kind of problem they solve; this makes them easier to use.

Classical TRIZ: Su-Field Analysis and the five classes of Standard Solutions

In Classical TRIZ there are two methods for understanding problems: Function Analysis and Su-Field Modelling. Both describe and define interactions: Function Analysis as subjects acting on objects and Su-Field Modelling as substances ('Su') interacting via a field ('Field'). Either can be used to access the 76 Standard Solutions, but the most common route in Classical TRIZ is described using Su-Fields. The Classical TRIZ Standard Solutions are organised into five classes:

- Class 1: Build (if missing or incomplete) and Destruct (if harmful) Su-Fields

- Class 2: Strengthen and Develop a Su-Field

- Class 3: System Transition and Evolution

- Class 4: Solutions for Detection and Measurement

- Class 5: Extra Helpers: How to apply the Standard Solutions

Su-Field Modelling and the Classical TRIZ Standard Solutions are powerful problem-solving tools and particularly useful when facing very complex problems of a highly technical nature. However, they are hard to use and understand and require quite in-depth teaching: Function Analysis and the Oxford TRIZ Standard Solutions are much easier to pick up and use, especially for people new to TRIZ, and still deliver powerful results.

There are many excellent books that go into more depth on Su-Fields and the Classical TRIZ Standard Solutions: Victor Fey's books are very good, and I can also recommend Gordon Cameron's *TRIZICS* and Karen Gadd's *TRIZ for Engineers* as they both also cover Function Analysis and are easy to read.

The 76 Solutions in Oxford TRIZ are arranged to solve three kinds of problem:

- ✔ **Dealing with Harm – 24 solutions**
- ✔ **Overcoming Insufficiency – 35 solutions**
- ✔ **Difficulties with Measuring and Detecting – 17 solutions**

The Standard Solutions are very 'listy': you have a lot more suggestions for new solutions than when you're looking at, for example, contradictions. When you're solving contradictions, three or four Inventive Principles may be suggested; when you're looking at harms, 24 suggestions are available. Some solutions will be a less obvious fit in your specific situation than others, so when using the Standard Solutions, don't get bogged down if you're struggling with particular suggestions; keep moving and you'll find that the solutions start overlapping and inter-connecting.

Dealing with Harmful Actions

Harms are one of the commonest types of problem. A *harm* is any output that you don't want. The full list of solutions for dealing with harms is available in Appendix E.

Harms can be things that are actively harmful in your system or simply outputs that aren't useful to you, such as heat from a light bulb or more programme options on a washing machine than you need.

The point is that output requires an input – and if you don't want the output, and you're paying for it in some way, eliminating that output will probably reduce your inputs and costs. For this reason, you treat excessive useful actions in the same way as harmful actions: although the outcome of providing more than you want may be good or neutral, you still need to know if you're getting too much of it. You can apply the Standard Solutions for dealing with harms to reduce the excessive useful action to a level at which it's just useful. You also capture risks as harms, even if they don't always happen: if they're a possibility you should capture them as problems that you can then deal with.

Components in your system are useful but also bring problems – an example is supermarket carrier bags. The bags provide useful actions – allowing you to pack and then carry your shopping home – but also a number of harmful actions. These include damaging the environment when they decompose, forming debris in the oceans and harming animals if they accidentally ingest them. And costing you 5p, of course! When you start trying to deal with the problems, you tackle one problem at a time. That's why you break down your system into all the different interactions: you want to improve your system without losing all the good things.

The best way to deal with a harm is to remove whatever's causing it – in TRIZ this is called Trimming. However, the component that's causing the problems may also have some useful actions: the power of the Trimming Rules is that they guide you to remove the component but without losing any of its useful actions.

The Trimming Rules are a list of questions you ask to see if you can find another way of getting the useful actions a component delivers – without the component:

- ✔ **Do we need its useful actions?** If no, trim it.

- ✔ **Could the object perform the useful actions itself?** If yes, trim the subject.

- ✔ **Could another component perform the useful actions?** If yes, trim it.

- ✔ **Could a resource perform the useful actions?** If yes, trim it.

- ✔ **Could we trim the component after its useful actions?** If yes, trim it.

- ✔ **Could we trim any harmful parts of the component?** If yes, trim it.

Sometimes the constraints of your situation don't allow you to explore Trimming. When this is the case, you can apply other Standard Solutions for dealing with harm: those which tackle just the harmful actions, leaving your useful actions untouched. This is such a powerful tool for problem solving that it's covered separately in Chapter 14.

You can also deal with harms in three other ways:

- ✔ **Block the harm:** Stop the harm having an effect.

- ✔ **Turn the harm into a good:** Find a way to get a benefit from the harm.

- ✔ **Correct the harm:** Accept that the harm will happen, then correct its effects afterwards.

Multiple solutions may be suggested for each of these, but it's worth remembering the broad categories, as they alone can suggest solutions to everyday problems. From the example in Figure 13-1 of a colleague annoying you, you can think of simple solutions from these categories:

- ✔ **Block the harm:** Counteract by being super-nice or 'love-bombing' the person. Perhaps he's being annoying because he's anxious or worried; even if not, your behaviour will be so confusing he may abandon trying to wind you up and leave you alone.

✔ **Turn the harm into a good:** Treat it as a learning experience. Observe all the annoying behaviour and make a note to yourself to never do this to another person.

✔ **Correct the harm:** Accept it's going to happen, and correct the effects of the annoyance on yourself by going for a long walk, eating a bar of chocolate or blowing off some steam to a friend later.

Improving Insufficient Actions

Very often you have something you want but you don't get enough of it: how much you're paid was mentioned above, but other examples could be stopping a car in the rain or cardboard cartons protecting eggs in your grocery shopping. If this is the case, it's described as useful, but insufficient. In an ideal world you'd get more of it. Fortunately, you can apply the Standard Solutions for improving insufficiency.

If you're not certain whether you're getting enough of something good, mark it as insufficient because this prompts you to look for ways to improve it. You'll be encouraged to find even better solutions and to continually challenge yourself to improve your system. You can always decide later that no improvements are necessary, but at least you're identifying as many possible opportunities to improve as possible.

Sometimes, there are useful actions that you want but you don't currently have: these are missing actions. These are sometimes hard to include in a Function Analysis (see Chapter 12 for more on these), because when you're drawing a Function Map, you're describing the state of the situation as it currently is, rather than thinking about other things that you want. However, if they occur to you, include them, perhaps marking them slightly differently, such as with a different kind of dot in the arrow line. They can be treated in the same way as insufficient actions: they're just insufficient to the point of not currently existing!

Measuring and Detecting

The requirement to measure or detect something is a particular type of problem. If you're looking to measure or detect a particular parameter, for example, temperature, you can use the Effects Database to find specific ways the world has found to measure temperature, weight or volume.

However, sometimes your circumstances or the nature of the problem make it very hard to measure or detect; Standard Solutions offer you inventive suggestions of ways around the measurement or detection.

There are three broad categories of suggestions:

✔ **Change the system so no need exists to measure or detect.**

This means understanding what the impact of the measurement or detection will be and focusing your attention on achieving this outcome itself; this can mean making part of your system self-serving, bypassing any need to measure or detect. For example, if you're measuring something so you can take an action when it goes over a certain level, make the system self-adjusting or with an auto cut-off.

✔ **Measure a copy.**

For example, if you want to measure something that's dangerous to get close to or moves very fast, such as a snake, you can take a picture and measure from that.

✔ **Introduce something that generates a field you can measure.**

For example, the outline of certain internal digestive organs is difficult to distinguish in an X-ray. To make them easier to see, patients can ingest barium, which shows up very brightly on X-ray. The barium coats the organs and the radiologist can see the barium very clearly, which tells them what the internal organs look like.

Use these approaches when for some reason it's hard to measure something. If you just need to measure something and you know how it can be done, you don't have a difficult problem – you can just do it. For example, if you want to measure your weight, you just stand on some scales: the problem isn't hard, so you don't need an inventive solution.

You need the Standard Solutions for Measurement and Detection when measuring or detecting something is challenging, as this situation requires you to be inventive. For example, monitoring someone's brain activity is hard to do directly; even if you could see through someone's skull, you can't map which parts of the brain are in use with the naked eye. You can measure electrical activity in the brain with an EEG machine, but this has very low spatial resolution. For this reason, clever measurement instruments have been developed that measure brain activity by proxy. Another brain-imaging method is fMRI (functional magnetic resonance imaging), which applies a strong magnetic field to the brain. Oxygen-rich blood responds differently to oxygen-poor blood, and the fMRI can map the flow of oxygen-rich blood in the brain: when a part of the brain is more active, it needs more blood, so activity can be detected via blood flow rather than by trying to monitor the activity itself.

Keep on running

An example from the sport of running is monitoring times: professional marathon runners typically monitor their performance by splitting the race into a number of sections or 'splits' and establishing targets to hit for each; Paula Radcliffe's exceptional performance in the 2003 London Marathon she attributes, however, to a 'no limits' philosophy. She said while she broadly aimed to run faster in the second half of the race, she didn't set herself targets to aim for but instead ran on 'feel'. As a result, she broke the previous world record with a time of 2 hours, 15 minutes and 25 seconds – a phenomenal achievement and a record that remains unbroken.

The point is that measuring isn't usually done as an outcome in itself but as a way of monitoring performance in order to change or modify it in some way. This approach isn't very TRIZzy, as it generates work (costing time and energy) that isn't directly related to the Prime Output, that is, the performance. Taking out the measurement and creating a self-system that doesn't require measurement is much more TRIZzy, as this approach directs all attention towards the Prime Output of your system: the one main thing you want to achieve.

The Standard Solutions for Measurement and Detection, like all TRIZ tools, prompt you to act like the smartest people and ask, 'What do I really want here?' This question directs you to consider more than perfecting the measurement system itself and is a bit like the joke about two men meeting a bear in the woods – the clever one knows that to survive he doesn't need to outrun the bear, just his friend.

The full list of Standard Solutions for Measurement and Detection is available in Appendix E, but it's worth starting with these three broad classes, of which the first is the most bold: remove the need to measure or detect. This isn't the way the world is going, as ever-more systems, particularly management systems, employ more and more detailed measurement. Quotas, targets and other measures are routinely used in all kinds of organisation – businesses, schools, hospitals, prisons, universities – in an attempt to monitor performance, so changes can be made if necessary to improve the performance. It would be much more inventive to change systems so the effort is put directly into the performance without any measurement required.

The *Prime Output* is what your system exists to deliver: its main purpose.

Radcliffe, whose marathon running strategy is mentioned in the preceding sidebar, said she didn't see the point of keeping targets as she wasn't going to slow down if she was doing better than expected. Her attitude is equally applicable in other fields: if you focus all your attention and energy on getting the performance you really want, you may do significantly better, partly because the energy and effort you put into measuring and monitoring your performance is redirected towards the outcome you want. Many professions are currently subjected to lots of performance measurement and management, such as nursing and teaching, which take time and energy away from their prime functions and towards administration. Greater trust may produce greater results (and more happy, productive people).

As with the other Standard Solutions, those for measuring are generated from engineering and scientific solutions, and are very useful when developing new technical solutions for measuring and detecting. The detail of the Standard Solutions often suggests very technical systems, for example one of the suggestions is using resonance or an object's resonant frequency.

However, there's no reason why these solutions can't be used for more general problem solving or management systems. Most organisations have systems for measuring all kinds of things: staff performance, sales figures, profit and loss and so on. The Standard Solutions, particularly in their broad classes, are useful for finding inventive solutions for all these measurement issues.

Applying the Standard Solutions

As soon as you have a well-formulated problem or series of problems, you can apply the Standard Solutions to generate inventive new solutions. We'll go through the steps you need to take to use the Standard Solutions, and apply them to a real-world problem.

Developing a well-formulated problem

Standard Solutions work best on SaOs: to apply the Standard Solutions most effectively, you must first develop a model of your problem as an SaO.

If you want to do some fast problem solving on a small problem, you can develop a simple SaO such as the one suggested in Figure 13-1 (earlier in this chapter), where the action of 'annoys' is a harm. This can be useful for some problem situations, and is the simplest unit of problem solving. You've described your situation as one interaction: a subject is performing an action on an object (annoying it – or, in this case, you!).

When you already know what the main cause of your problem is, starting with such a simple problem description can give you a fast-track to the Standard Solutions. However, I recommend completing a Function Analysis if you have time and your constraints allow it: the Function Analysis will show you much greater detail and may clearly reveal the real reasons behind the causes of the problem.

An example of Function Analysis, using the problem of noise from the ventilator scenario in Chapter 12, would be looking at how the ventilator works, its specific useful actions in terms of treating the patient and what it is about its design and functioning that makes it noisy, and then dealing with those technical problems one by one. This gives you many more opportunities to improve the ventilator. However, if you can't change the ventilator (for example, you're a nurse on the ward, not an engineer who makes ventilators), looking at problems at this high level can still give you access to useful, practical solutions.

Solving problems outlined in a Function Analysis

A *Function Analysis* is a description of a problem that's both very precise and very conceptual. This is because it's described in terms of functions. You can read much more about them in Chapter 12.

A *function* is the description of how an action changes an object. In an SaO, the function is the combination of the action and the object: the subject is the thing delivering the function.

SaOs are the building blocks of a Function Analysis. When you complete a Function Analysis, you're analysing each individual interaction in your system.

Identify, define and then categorise problems into three different kinds:

- ✔ Harmful actions
- ✔ Insufficient actions
- ✔ Excessive actions

You can then apply the Standard Solutions to solve the problems you've identified: one by one.

One of the values of Function Analysis is that it lets you see how all the components of your system interact, all the links and relationships, but it also gives you focus, because when you problem solve you tackle just one problem at a time. When you start applying the Standard Solutions to find new solutions, you do so just one SaO at a time.

You may have a chain of problems that connect or many that you want to tackle in total, but whatever the case, you only focus on one individual SaO at any time. This makes problem solving less intimidating because you don't have to grasp the whole situation and consider all the links throughout; you merely have to find solutions to the problem in front of you, which is much smaller and more manageable and well-defined. By working through your whole Function Map, you can be confident that you've covered all the problems and found all the potential solutions, but you've done so by working through a series of smaller tasks.

Generating innovative solutions

As soon as you have a completed Function Analysis (check out the previous section if you've not already read it), start finding innovative solutions to your problems!

When you take a real-world problem and model it in a conceptual way, develop an SaO or a series of SaOs you've built into a Function Map, you're stepping through your Prism of TRIZ. You then have a well-formulated problem to which you can look up the answer – in the Standard Solutions. The next bit is the most fun: now you have to take the conceptual solution and work out how it could be made a reality. This is where your experience and expertise come in, and your creativity is stimulated, because you're not starting with a blank page but a solution suggestion: a very conceptual solution that you have to turn into a specific idea.

Organising the Oxford TRIZ Standard Solutions

The Standard Solutions have been organised into three classes, and are numbered in a logical way. Each Standard Solution has both one or two letters and two numbers: the letter tells you the class of the Standard Solution; the first number tells you the class of Standard Solution, the second number identifies the specific Standard Solution.

Dealing with Harm

H1 = Trim out the harm (6 solutions, so H1.1, H1.2 and so on until H1.6)

H2 = Block the harm (11 solutions)

H3 = Turn harm into good (4 solutions)

H4 = Correct the harm afterwards (3 solutions)

Overcoming Insufficiency

i1 = Add something to the subject or object (7 solutions)

i2 = Evolve the subject and object (10 solutions)

i.a. = Improve the action (18 solutions)

Difficulties with Measuring or Detecting

M1 = Indirect methods (3 solutions)

M2 = Add something (4 solutions)

M3 = Enhance measurement with fields (3 solutions)

M4 = Use additives with fields (5 solutions)

M5 = Evolve the measurement system (2 solutions)

All of the Standard Solutions work by first giving you a very radical suggestion, followed by successively less radical suggestions. This is the typical TRIZ approach: you start with trying to make the biggest change possible, and only if it's impossible do you move on to the smaller changes.

The steps for applying the Standard Solutions are:

1. **Define an SaO or complete the Function Analysis (see Chapter 12 for more details).**

   If you're starting with an SaO, go straight to Step 5.

2. **Pick a place to start problem solving.**

   Good places to start are:

   - Near the target or Prime Function of your system: the main thing your system is designed to deliver.

   - Components with lots of problems: spotting where many harms, insufficiencies and excessive actions are coming from is easy because the different style lines stand out clearly.

   - Components that have low Ideality: if they deliver few useful actions, and have lots of costs and harms associated with them, they'll be ripe for improvement.

3. **Start by trying to trim components (see Chapter 14).**

   After that, follow the steps below.

4. **Pick one action that's harmful, insufficient or excessive that you want to tackle.**

   Read or write out the SaO again.

5. **Look up the relevant Standard Solutions for dealing with the problem.**

   Read them out loud (this is really helpful), replacing the Subject–action–Object in the Standard Solution with the one in your SaO.

   - Try to think of new ideas and solutions suggested by the Standard Solution.

   - Try multiple Standard Solutions, capturing all your ideas as you go.

   - When you've generated many new solutions, see if you can combine them to make a better system. The best solutions often come from combining many solutions: try to do this as much as possible.

   - If you have time, repeat this process with your new, improved system.

 The solutions shown in this section are from just a small number of Standard Solutions. There are many more to play with and I suggest you check out Appendix E and see if you can generate any better solutions to the ones suggested here!

It's worth noting that the Standard Solutions often suggest changing things outside the system: using the environment (the *super-system* – everything that's outside your system) and mobilising resources or even changing things around the system. This approach may seem outside the boundaries of the problem you started with, but it's worth considering.

TRIZ constantly challenges and pushes at constraints: it's very easy to keep your head down and focus on the problem you were brought in to deal with, but very inventive solutions often bring in clever use of the things around you, and the Standard Solutions regularly prompt you to look beyond your system. The Standard Solutions will not only keep you focused on developing clever solutions to your well-defined problems, but will also help you break your psychological inertia and challenge what you thought was possible.

Let's step through examples of using the Standard Solutions to generate new solutions for the rest of the chapter. Are you sitting comfortably?

Tackling a real problem

Let's look at a real-life scenario, and consider how Standard Solutions could help us generate useful solutions for dealing with noise in intensive care units (ICUs), the problem described in Chapter 12.

Intensive care units are noisy places: very ill patients are attached to monitors (equipment that measures heart rate, blood pressure, breathing and so on), which sound a noisy alarm if a patient's vital signs change in a worrying way, requiring a doctor or nurse to attend to the patient, check him and turn the alarm off. Many other sources of noise also exist: ventilators, telephones, radios, televisions and also staff talking to each other around the patients' beds and across the ward. All this noise can disrupt patients' sleep, causing them stress, contributing to delirium and resulting in poor health outcomes. Let's consider how some of the Standard Solutions could generate useful solutions to these problems (the full Function Map for this situation can be found in Chapter 12):

Step 1: Complete Function Analysis (done in Chapter 12).

Step 2: Recognise that the two largest causes of disruptive noise are monitors and staff talking.

Step 3: Avoid removing monitors or staff completely, as doing so would involve too large a change (both medical and hospital practice).

Step 4: Consider the noise created by monitors and staff talking, as shown in Figure 13-2.

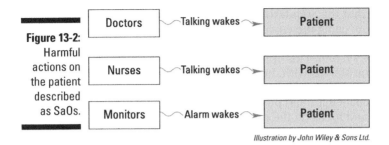

Illustration by John Wiley & Sons Ltd.

Figure 13-2: Harmful actions on the patient described as SaOs.

Step 5: Apply Standard Solutions.

H2.1 Counteract the noise with an opposing field that neutralises the noise.

This suggests using other sounds to block out the noise: white noise or ocean sounds (improves sleep by about 40 per cent). A resultant benefit is increasing patient confidentiality (it's harder to hear staff talking about other patients).

H2.3 Change the zone and/or duration of the noise to decrease its effects.

This suggests reducing the amount of loud talking that occurs between doctors and nurses by tackling either where or when the conversations happen. Some neonatal ICUs remove the patients to private rooms for treatment, so that the necessary dialogue between members of staff doesn't disturb other patients. Some ICUs have also tried implementing quiet times during the day and night, when no visitors are allowed, all radios and TVs are switched off and staff are asked to keep their voices low.

H2.4.1 Insulate from the noise by introducing a new component or substance.

This suggests adding something to insulate from noise: ear muffs and ear plugs have been tried (improves sleep by about 25 per cent).

H2.7 Protect part of the system from noise.

If you can't do everything you want, protect parts, starting with the most important. In this case, it suggests separating the most critically ill patients (who require more staff intervention) from the those recovering from surgery, so the latter are in a quieter zone.

H2.8　Reduce the noise by using a weaker action and enhancing it only where required.

Some hospitals have created quiet zones around patients by asking staff to keep their voices down and creating sound-proofed clear offices in the middle of the ward where the nurses and doctors can talk freely.

H2.9　Use sub-systems/details of components to stop the noise.

This suggests creating devices with other types of alarm, for example lights, or using a noise that can somehow be heard only by doctors and nurses. Another idea is to find quieter ventilators.

H2.10　Use super-systems/the environment to stop the noise.

This suggests using soundproofing on the windows and noise-absorbing materials in the walls, and lowering ceilings to absorb ambient noise and reverberations of unavoid-able noise.

Step 6: Apply as many of these solutions as possible to create a much quieter environment.

Improving a solution

Sometimes you've got something good but it's just not good enough. And being a TRIZ problem solver, you want to make it better! The Standard Solutions contain 35 suggestions for improving any insufficient useful actions, which you can apply to improve the humble toilet brush. I figure if I can explain this using a toilet brush as an example, you should be able to apply it to any problem!

A toilet brush is something most people are familiar with, and the design hasn't significantly changed for some time. Let's imagine you're a toilet brush designer, and want to create a better brush:

Step 1: Complete a Function Analysis of the situation, as shown in Figure 13-3.

Step 2: Pick a place to start problem solving. As the Prime Function of this system is 'brush – moves – debris', start with the brush.

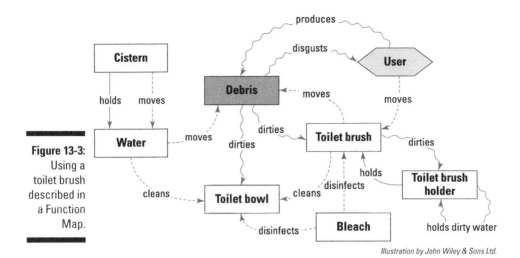

Figure 13-3:
Using a toilet brush described in a Function Map.

Illustration by John Wiley & Sons Ltd.

Step 3:   Trim the subject: as a toilet brush designer, you may not want to start by removing your system completely (although this will generate very deep trimming, as described in Chapter 14). In this instance, move on to the other Standard Solutions.

Step 4:   Pick an SaO to improve. As the Prime Function in this example is insufficient, this is the most obvious place to start. Look to improve 'brush – moves – debris'.

Step 5:   Look up the relevant Standard Solutions, replacing the SaO in the Standard Solutions with those in your SaO.

   i1.1   Add something to or inside the toilet brush or debris.

   This could suggest making the toilet brush handle hollow, and adding bleach or some other detergent inside the brush that's automatically dispensed in use.

   i1.3   Use the external environment to enhance/provide the function of 'move debris'.

   This suggests changing the toilet bowl in some way (potentially outside of your constraints, but worth considering). Encourage the debris to move by vibrating the toilet bowl, or adding a 'cleaning' setting to the toilet, which dispenses a small amount of water from the cistern to aid the cleaning process. You could make a non-stick toilet bowl.

i1.4   Change or add something to the environment/surroundings of the toilet brush and debris.

Perhaps you could add something to the water in the cistern that either makes the toilet bowl less easy to stick to or aids cleaning by loosening the debris.

i2.1   Segment the toilet brush or debris: increase the degree of fragmentation or divide into smaller units.

You could fragment the bristles: either have many more thin bristles to improve the cleaning action or have micro-bristles on the ends of each bristle.

i2.2   Introduce voids, fields, air, bubbles, foam and so on into the toilet brush or debris.

This could suggest adding foam: perhaps you create a foaming bleach or some other detergent to increase the useful action.

i2.10   Improve a system by changing components/substances to deliver exactly what's needed in time and/or space.

Perhaps you could develop a disposable toilet brush: one of the problems identified in the Function Map was that the dirty toilet brush sits in a holder. If the bristles were only there when needed (that is, when cleaning the toilet), that harm would be eliminated.

i.a.2   Add another action from existing components.

The action is 'move', which suggests adding another action somehow. Perhaps in addition to the mechanical action, you could add a water jet (water is available). You could imagine a toilet brush that sprays water in addition to using the bristles.

i.a.4   Change from a uniform action (move) to an action with predetermined patterns.

This suggests that, rather than a single movement, you vibrate the bristles. An analogous system is the toothbrush (apologies for the analogy in this setting!) – perhaps you could invent a vibrating toilet brush or even a sonic toilet brush!

Step 6: Combine the ideas.

You could imagine putting a number of these ideas together: creating a vibrating toilet brush that also sprays either water or detergent from the handle and has smaller bristles.

Chapter 14

Trimming for Elegant, Low-Cost Solutions

. . . .

In This Chapter
▶ Getting more with less
▶ Developing elegant solutions
▶ Creating strong intellectual property

. . . .

Generally, when you try to make something better, you do so by adding something else: *Trimming* is the TRIZ way of making your system better by removing things while keeping all the useful actions.

While the logic behind why Trimming works may seem complex, in practice the steps are easy. You need to start with a system that you fully understand and complete a good Function Map: all you need to do then is pick a component, and question whether you can remove it using the prompts in the Trimming Rules. You'll find that Trimming gets easier the more you remove – and sometimes removing one troublesome component opens up the opportunity to remove a handful of others.

In this chapter I give you my hints and tips for Trimming your way to success.

Making Things Better and Cheaper

Trimming is radically different to traditional problem solving – you start with the idea that you're going to improve your system by taking things away, and get more of what you want with less. When you follow the Trimming Rules, you systematically remove components while retaining all their necessary outcomes.

Improving systems by removing stuff

All of the TRIZ tools and processes are geared towards improving Ideality.

Ideality is the TRIZ measure of how good something is. Ideality is the ratio between all the benefits that you want and all the inputs required to create it (the costs) and the outcomes you don't want (the harms). You can think of it as a simple equation:

Ideality = Benefits / Costs + Harms

Systems improve as they get more benefits with fewer costs and fewer harms. Trimming improves systems by focusing on reducing the bottom half of the Ideality equation: if you can remove parts of your system but somehow still get all the same or even more benefits, then you reduce costs without making your system worse, and may have made it better. And you do so by focusing on functions: your costs are the inputs required to deliver functions, which give you benefits and harms, as shown in Figure 14-1 (you'll find a similar-looking diagram in Chapter 5, but see if you can spot the differences!).

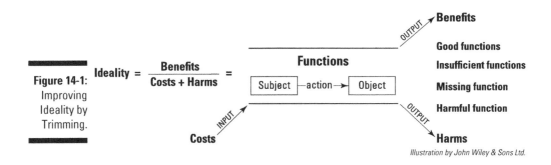

Figure 14-1: Improving Ideality by Trimming.

Illustration by John Wiley & Sons Ltd.

When you trim, you remove components from your system, losing all their costs and all their harmful functions but always aiming to keep their useful functions – which are delivering benefits.

One of the reasons Trimming is such a powerful problem-solving tool is that it provides you with a structured approach to dramatically altering your system, by getting rid of the most troublesome and expensive parts. Trimming encourages you to deal with problems by removing their causes completely, rather than trying to make small changes to manage the subsequent downsides. It's the first suggestion for dealing with harms in the Standard Solutions because it's the most radical (Chapter 13 covers the less radical Standard Solutions).

Trimming yourself Lean

Trimming encourages the development of very lean and efficient systems because removing as much stuff as possible from your system often means you create more sustainable designs. You use fewer materials to produce your system and more efficient systems require less energy. James Dyson's bagless vacuum cleaner is a good example of a trimmed system: the bag is trimmed out but its useful action (separating dust from air) is still delivered albeit via a cyclone system. As well as fewer input costs (no need to buy bags for your vacuum cleaner), fewer harms result too (no bags full of dust disposed of in landfill sites).

Trimming fits very well with other problem-solving approaches, such as Lean manufacturing, which look for ways to make products and processes more efficient by reducing time and resources required and reducing waste. Trimming Rules are very popular with Lean professionals for this reason: when Lean tells you that something should be removed (the 'what'), the Trimming Rules suggest a number of ways in which this can be done (the 'how'). Trimming works by transferring responsibility from the component you're removing to other parts of your system, and a number of strategies exist for doing so, as shown in Figure 14-2.

Before you can start Trimming, you need to complete a Function Map (see Chapter 12): you then have a clear idea of all the components in your system and can work out how to remove some of them by transferring their useful actions to other parts of the system. With a completed Function Map you can allay the fears of colleagues or employees whose response to a proposed change is, 'We can't remove that; we need it!' The Function Map shows them that its actions will be carried out elsewhere in the system. When you go through the Trimming Rules, you can then work out alternative ways of getting all the things you need, thus delivering innovation without risk.

What's useful about this process is that it's a systematic way of breaking the perceived connection between the functions you want and the current system that's delivering those functions. Before you can start transferring responsibility within your system, you first need to complete a Function Analysis of that system: you start Trimming from a Function Map.

TRIZ *Function Analysis* is the process of understanding and capturing the functions (both good and bad) of your system. A *Function Map* is the picture that you create by the end of this process (Chapter 12 takes you through the steps).

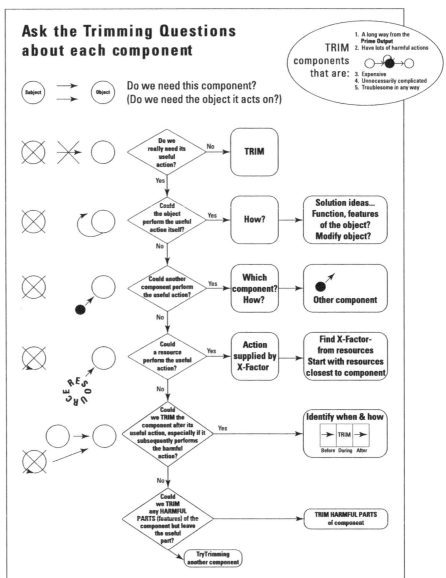

Figure 14-2:
The Trimming Rules.

Source: Oxford Creativity

By completing this process you ensure that you really understand how your system works because you capture all of its functions. This allows you to understand how all the different parts of your system work together, which bits are doing useful things and where the problems are located.

A Function Map captures four kinds of actions: useful actions, insufficient useful actions, harmful actions and excessive actions. When you remove a component, it no longer performs any of the harmful actions (which is good) but you also lose its useful actions (which is bad). What makes Trimming clever is that it makes you look for ways to remove a component but keep all its useful actions, which is why you need to complete a Function Analysis first. You need to understand all the useful actions each component delivers, so that when you remove it you can ensure you don't lose anything useful the component delivers. You need to transfer even insufficient and excessive useful actions, because in the process you may find ways to improve the insufficient actions and reduce the excessive actions so that they're simply useful.

A Function Map comprises a series of Subject–action–Object relationships. Trimming involves removing the Subject of any one of these interactions and finding something else that will deliver the action–Object (the function). In this way, you transfer the functionality of each component to something else – one action at a time. If a component is delivering three useful actions, you go through the Trimming Rules for each, generating as many ideas as possible.

Clever cost-cutting: Doing more with less

Trimming provides you with systematic prompts to get all the things you need. It enables you to cut costs intelligently by understanding the functions you need and finding new ways to deliver them.

Anyone can cut costs and reduce benefits at the same time. What's really clever is cutting costs without losing anything useful, and the Trimming Rules provide you with a systematic method not just for doing that but also simultaneously increasing benefits. This is one of the reasons why TRIZ innovation is so important, especially when times are tough: when budgets are slashed and people made redundant, that's not the time to continue with business as usual. During lean times, you're often pressured to deliver the same output but with fewer resources. You thus need to find a way of doing at least all the things you used to (more if possible), but with less. Commercial pressures may affect mature products and promising new products and concepts – both are vulnerable to cost-cutting initiatives and the latter may fail to reach the market if they can't be produced cheaply enough. While traditional methods can assist with incremental cost-cutting, when you're expected to deliver significant cost reductions, you need TRIZ's Trimming Rules.

By understanding your system – whether a jet engine or department of people – in terms of its functions, you can model the useful outputs every part of it delivers. Rather than focusing on the details of how your system currently operates, modelling its functionality allows you to uncover problems *and* identify all the good things currently in existence. You can then consider whether you can change your system by reducing it in size: removing components one at a time but without losing any useful outputs. When you've mapped how everything in your system works, you've captured all the things your system currently delivers. These are the things you really want, and the components of your system are just one way of gaining them. Knowing this, you can begin to trim and reduce costs without making anything else worse. Indeed, you may actually produce benefits: reducing the weight of a car, for example, not only decreases the materials needed but can also increase acceleration; trimming out unnecessary meetings can increase productivity because that time can be spent doing other work.

The Trimming Rules prompt you to find other (and often better) ways of delivering all those useful functions. They also point you to specific places to discover those other ways. The Trimming Rules are also iterative: you don't stop when you've removed a single component; you keep Trimming until you've removed as many components as possible.

What often happens is that when one or two components can be removed, a whole new way of delivering what you want often becomes an option. At this point, you're not only finding new solutions and new ways of getting what you want, but also asking new questions that may uncover new opportunities. Don't stop at this point. Redraw your new system as a new Function Map and trim again. You may find new locations in which to use your system and new functions, potentially providing significantly more benefits as well as fewer costs and harms.

Shining a light on Trimming

Oxford Creativity's TRIZ learning workshops involve a Trimming exercise with a traditional, battery-powered torch. Starting with a system that contains 13 components, participants can often trim it down to a simple four-component system consisting of an LED bulb, battery, switch and housing. At this point, we encourage them to conduct another Function Analysis, draw a new Function Map and trim again. This second phase generally generates much more radical suggestions, such as a single-component system made of bio-luminescent particles that fluoresce on demand: a completely new system with new opportunities and new benefits (as well as new questions and new problems!).

Sometimes during the Trimming process people start 'smuggling' new components into the system. When something's been removed and a new way of doing things suggested, they may see something else that could also be delivered by the system – provided another component is surreptitiously added. That's okay. Add in the new component and its extra useful outputs as a means of capturing the new functions you want. Then try to trim the new component! You may end up with a trimmed system that gives you more of what you want rather than less – but with fewer components at the end of the process. Consider coffee shops: they often have milk delivered in floppy plastic bags rather than plastic bottles. The milk is then transferred into rigid, reusable plastic jugs that are easy to stand on the counter and pour from, and to store in the fridge. This system is clearly the result of an attempt to trim in space. Floppy bags take up much less room than plastic bottles but are difficult to store when opened. To make this floppy bag system work, another component has been smuggled in: the new plastic jug. One of the functions required of a bottle – rigidity – has been transferred to this new component.

Using resources wisely

In order to trim effectively, be aware of all your available resources. One of the Trimming Rules suggests transferring responsibility for an action to a resource, and in order to do so, you first have to be aware of all the resources that exist! Listing your resources before commencing Trimming is a good idea. While you can just jot them down on a piece of paper, consider capturing all your resources in your Function Analysis. Next to each component, list the resources available for that component. When you start trimming and want to transfer the useful action to a resource, you can then begin with resources close to the component you're trimming.

Chapter 5 covers in detail how to capture resources; here, we simply point out that the most useful prompts are

- Super-system and environment
- System features
- Sub-system and component resources
- Time
- Space
- Harms and waste
- Knowledge, experience and feelings

Before you start Trimming, list all your resources, so that when you come to that step in the Trimming Rules, you can keep thinking fast.

Creating Elegant Solutions

Applying the Trimming Rules enables you to generate elegant solutions, whereby you provide the same (or more) benefits with fewer costs. Following the Trimming Rules helps you see not only what you could remove but also how you can remove it without losing any benefits.

Applying the Trimming Rules

Here's a concrete example to help you see how to apply Trimming Rules.

Many hotels provide a mini-bar for their guests. If the hotel's information literature hasn't already informed you, the hum of the fridge often notifies you of its presence – and as soon as you know it's there, you're tempted to look inside. A cool, refreshing drink may be just the thing after a long, tiring journey! However, one look at the high prices may put you off; the appealing little bottles and cans are often much more expensive than their equivalents in the hotel bar. But, you reason with yourself, they do have the virtue of being instantly available in your room, day and night.

The Prime Function of this system is to ensure that drinks are available to refresh the guest. The mini-bar delivers the Prime Function by providing refrigerated drinks in the hotel room. The drinks are expensive, which is good for the hotel because they generate income, but bad for the guest because it impoverishes her. The secondary functions of the mini-bar are to inform the guests that drinks are available and to tempt them to buy them.

The mini-bar also, however, requires inputs. It needs electricity, and its contents must be replenished daily by hotel staff and charged to the right bill How can you improve this system by making it simpler but without losing all the benefits?

The Function Map in Figure 14-3 gives you some ideas.

You need to follow the Trimming Rules to improve this system. First, you need to select a place to start Trimming, for example with:

✔ Components that create lots of problems (generate harms)

✔ Components that have low Ideality: they deliver few useful actions and are associated with lots of costs and harms

✔ The Prime Function: for more incremental problem solving, start far away from the Prime Function; for more radical problem solving, start close to the Prime Function

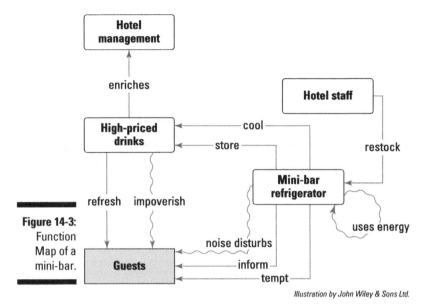

Figure 14-3:
Function
Map of a
mini-bar.

Illustration by John Wiley & Sons Ltd.

The mini-bar seems to have two problems (it creates noise and requires energy input) *and* it doesn't deliver the Prime Function. The mini-bar is therefore a good place to start.

Before trimming you need to capture all possible resources:

- ✔ **Super-system and environment:** Corridors, lifts, bar, reception, staff, kitchen, other local businesses/bars

- ✔ **System features:** Hotel room

- ✔ **Sub-system and component resources:** Cupboards, air-conditioning unit, TV, hotel information brochure

- ✔ **Time:** When booking, checking in, walking to room from reception

- ✔ **Space:** Inside and outside room

- ✔ **Harms and waste:** Waste from drink storage, expense, noise from and energy use of refrigerator

- ✔ **Knowledge, experience and feelings:** Of staff and guests

You then need to list the useful actions the mini-bar delivers. The Function Map in Figure 14-3 suggests that the mini-bar provides four main useful actions: it cools and stores high-priced drinks, and informs and tempts guests. You now work through the Trimming Rules for each of these four actions, asking yourself whether these useful actions can be delivered in other ways. When you find an answer, don't stop; simply capture it and move on to the next suggestion. If you were tackling this problem in real life, you'd work through each of the four actions in turn; here, however, Table 14-1 presents suggestions for all four, organised according to which Trimming Rule is being applied.

Table 14-1	The Trimming Rules in Action			
	Mini-Bar Refrigerator Stores Drinks	*Mini-Bar Refrigerator Cools Drinks*	*Mini-Bar Refrigerator Informs Guests*	*Mini-Bar Refrigerator Tempts Guests*
Do we need its useful action?	No – use cupboard	No – only provide drinks that are served at ambient temperature, for example, red wine, water	Yes	Yes
Could the object perform the useful action itself?	Extend drink packaging to make a self-contained pack of many drinks	Self-cooling drink cans	Drinks on sideboard/in plain view	Drinks on sideboard/in plain view
Could another component perform the useful action? The only other components in this system are the hotel management and hotel staff	Check-in desk stores drinks – guests can purchase when checking in No mini-bar: offer free room service instead (drinks stored and cooled centrally)	Hotel staff bring ice on request	When helping with luggage, staff can offer to bring guests a drink/inform of the drinks service	Guests can pre-order drinks when booking room Offer all-inclusive package to include drinks for a fixed charge: guests pick up pack on check-in
Could a resource perform the useful action?	Use wardrobe/cupboard to store drinks	Use air-conditioning unit to cool drinks	Welcome screen on TV advertises drinks/drink availability	Browse and order available drinks via the TV (delivered by room service)

	Mini-Bar Refrigerator Stores Drinks	*Mini-Bar Refrigerator Cools Drinks*	*Mini-Bar Refrigerator Informs Guests*	*Mini-Bar Refrigerator Tempts Guests*
Could we trim the component after its useful action?	Disposable cardboard packaging for drinks	Offer chilled drinks in a free/ disposable cool bag from reception	Leaflets advertising drinks and prices left on bed	Leaflets have daily special offers, for example, today's cocktail
Can we trim part of the component?	Offer vending machines – one per floor – instead of mini-bars in each room Smaller fridges for which guests preorder specific drinks	Offer ambient temperature drinks and ice machines on each floor	Adverts for drinks in lifts	One free drink available in room – with details on how to purchase others

Checking that you've drawn a perfect Function Map

You can't always guarantee that you've developed the 'perfect' Function Map of any system, as the diagram that you create at the end will reflect your knowledge and experience (for this reason, you can only really create Function Analyses of systems that you know and fully understand the workings of). You also have to select where you draw the boundary: in the case of the mini-bar described in the 'Applying the Trimming Rules' section nearby, you could be considering the supply chain for the drinks, the forms that are filled in to inform hotel management how much has been consumed and so on.

However, the Function Analysis shown in Figure 14-3 probably has sufficient information to be useful and trigger sensible ideas. Amending and improving your Function Map as you go through problem solving is perfectly acceptable. Sometimes when you try to trim something, you realise that you've missed out a useful action. Again, that's fine; add it in and keep problem solving. Completing a Function Analysis will require a certain amount of judgement; its purpose is to help you understand how your system really works, to focus on its functionality and outcomes, so that you can apply problem-solving tools such as the Trimming Rules to improve it. Your Function Map doesn't need to win awards or be objectively perfect: it needs to be useful!

Following the process and going off-piste

As with many of the TRIZ tools, the Trimming Rules provide you with a clear set of steps that should be followed in a certain order. However, when you start generating ideas, you may well find yourself taking off in interesting directions that may not seem to be related to the task in hand.

The Trimming Rules aim to make you think of other ways to get what you want. So, if you discover a rich seam of new ideas that lead you in interesting, useful directions, by all means follow them, as long as you're continuing to think of new solutions and generate interesting suggestions and points for discussion.

This seam will naturally begin to yield diminishing returns and eventually you will run out of steam: at this point you need to return to the process. Following the steps of the Trimming Rules ensures that you cover every possible solution. If you think of solutions that don't appear to be related to the TRIZ task at hand, that's fine; don't waste energy trying to reverse fit them into the process, just keep going and return to following the steps.

Going off-piste for short periods of time is fine if you're usefully coming up with new solutions. However, always return to the process to make sure you don't stop prematurely. Just because you've already thought of good ideas doesn't mean you won't think of better ones – allow yourself to be surprised! Following the process means that you can be confident that you've looked for every possible solution.

Trimming to Infinity and Beyond

If you're looking for highly innovative concepts, then trim, trim and trim again.

When do you stop Trimming? When you've gone too far. Knowing how much is enough is hard to gauge, so a good rule of thumb is to keep Trimming until you've taken too much out. All of a sudden you're losing useful functionality and your system stops working. At that point, take a step back.

Trimming starts as just a thinking exercise: when you're Trimming you're imagining new ways in which your system could work and developing conceptual solutions. This thinking time is (relatively) cheap and you'll recognise within a few minutes that removing another component makes everything go wrong. At that point you'll know that you've pushed yourself – and your system – as far as it can go.

You keep going because the more Trimming you do, the more innovative your solutions will be. Figure 14-4 graphically demonstrates the relationship between the amount of Trimming you do and the innovativeness of the final system you develop.

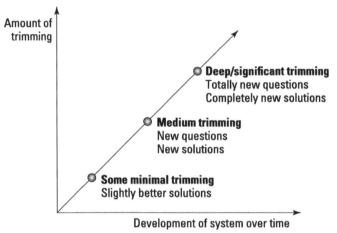

Figure 14-4: Trimming and system development.

Illustration by John Wiley & Sons Ltd.

When you trim out one or two components far away from your Prime Function, you create a slightly better system that has some minor improvements. Trim out more components that are more closely related to your Prime Function and you get totally new solutions that ask new questions about how your system can work and what it can offer. When you reach the very deep Trimming stage – for example, you've already removed the component that used to deliver your Prime Function or re-trimmed an already trimmed system – you can ask completely different questions regarding how your system operates and potentially consider new technology offering new opportunities. Ultimately, you'll develop a much better system.

An example of deep trimming is Internet grocery shopping, which has trimmed out the actual, physical shop you visit. The functions required to perform a grocery shop are now separated in time: you create your list online, and select the food you want while you are still at home – the actual delivery and putting away of the food is much later. This provides the additional benefits that you can select and choose your food very far in advance, both for holiday periods (which often require a lot of other work as well) and also at your convenience (in the middle of the night, while at home on the sofa watching TV). The cost of your time is also reduced because you don't need to travel to and around the shop, and your time and energy required to do the shop are reduced because you don't need to pick up items, put them in your trolley, then on the conveyor, pack them, load them in the car and carry them to your house: all these steps are performed by someone else. The harms of the hassle and stress of going round a busy supermarket (especially with small children!) are removed. There are also additional benefits available, such as potentially a wider range of products; certain household essentials can easily be ordered every week automatically; and you also stick to buying just what you need – without being side-tracked by appealing supermarket displays.

Trimming to Create Strong Intellectual Property

Trimming is a very powerful tool for ensuring that what you create is protected by intellectual property rights. Applying the Trimming Rules can help you avoid accidentally infringing someone else's patent or strengthen your own, and find freedom to operate for innovative new inventions.

Getting round someone else's patent

If you're seeking radically new concepts, you need to relentlessly trim. You should also actively look for ways in which to smuggle in new components – which will deliver new, but currently missing, functions. Completing an Ideal Outcome (go to Chapter 9 for details) will also help you spot benefits that your existing system doesn't currently deliver. And translating those benefits into functions will identify the action–Object you're looking for. You can then consider solutions that will provide that action–Object (using the Effects Database covered in Chapter 6). Even better, you can think of a solution and then trim it out! Ultimately, you'll end up with a very strong concept that's profoundly different to the system you started out with in terms of functionality and design. When you're Trimming to produce radically new concepts, you need to start with components that are closest to the Prime Function. Removing these components will prompt you to consider drastically new ideas and ways of doing things.

When people talk about patent circumvention, they usually mean that they're looking for a small change to make to a product so that their invention doesn't infringe an existing patent. However, in my experience, patent issues usually arise when one person discovers that someone else's patent is blocking something she wants to do.

One approach to patent issues (that's less vulnerable to the decisions of the courts) is to come up with a new concept that's so fundamentally different in its operation that it no longer bears any resemblance to the original patent. This is why TRIZ Function Analysis followed by Trimming is such a powerful approach to invention: even though you start with a real, existing system, by the time you've removed multiple components, you've often come up with radically different and innovative concepts. Your Function Map has helped you see all the good things that someone else's existing system delivers but you've developed, improved and simplified it by removing components while simultaneously keeping all their useful actions. You have also created an invention that is uniquely yours – and doesn't infringe the original patent.

Developing your own patent further

The same processes as described in the preceding section can be used to develop your own patent further, but you'll probably want to start with more incremental solutions. To generate improvements in your existing patent, look to trim out components far away from your Prime Function. These will generate small solutions that leave your basic system intact but can be added as dependent claims to your patents.

Also follow the steps discussed earlier in this chapter to add in missing functions with new subjects, and then trim them out to generate further developments in your patent. In addition, complete all the steps suggested in the preceding section for getting around someone else's patent because, in doing so, you're adopting the mindset of a competitor – but before your competitors have even seen your patent. This allows you to predict many (even better, all!) of the ways in which other people may try to circumvent your patent when it's published. You can then include these as other claims; even independent claims if the ideas you come up with are radically different to your original system. Trimming allows you to predict how your system will develop in the future; what you choose to do with this information depends on your patent strategy.

If you're planning to develop your own patent further, consult the Trends described in Chapter 4. Trimming and the Trends are the two most powerful tools for developing and protecting intellectual property.

Redrawing any new system you develop and trying to trim it again is always worthwhile. At the very least, apply the Standard Solutions (which you can read about in Chapter 13) to deal with any new potential problems. You'll generate a range of solutions that will allow you to plan the future for your system – and its next generation.

Part V
The Part of Tens

Head to www.dummies.com/extras/triz for a free article that lists ten online resources for finding out more about TRIZ.

In this part . . .

- ✔ Avoid ten common problems when tackling real-world problems.

- ✔ Use TRIZ effectively and apply the principles so they become a habit.

Chapter 15

Ten Pitfalls to Avoid

In This Chapter

▶ Avoiding a bad start with TRIZ

▶ Recognising the wrong kind of problems

▶ Spotting common problem-solving mistakes

*H*aving been round the block a few times with TRIZ, I see the same old mistakes time and again. This chapter describes the ten commonest pitfalls I try to steer my clients clear of, both when starting off with TRIZ and when tackling real-world problems.

Thinking TRIZ Doesn't Apply to You

If I received a pound every time someone told me that TRIZ wouldn't help him because the problems in his industry are completely unique, I'd be, well, loaded. However, believing that your problems are unlike anyone else's is flawed thinking – and not very TRIZzy.

People often ask to see case studies demonstrating TRIZ's application in their specific, often narrow, field before they believe it can be useful. The logic of TRIZ is that the same fundamental problems occur again and again across all industries and applications. One of TRIZ's greatest strengths is its fabulous pedigree of applying engineering and scientific research to create solutions. You can thus rest assured that you're actually *reapplying* tried and tested relevant proven solutions to solve your problems and meet your needs. This doesn't mean that TRIZ is only useful for engineering and scientific problems. Many of the TRIZ tools and approaches model the innovative thinking of creative problem solvers, which means you can re-create their patterns of thinking to deal with any kind of issue. More importantly, the logic of TRIZ tells you to lift your head and look beyond the detail of your own field. At the very heart of TRIZ is the philosophy that you can learn from the experience of others and that your situation is not unique. Someone else has seen something similar in the past, and parallels will exist elsewhere if you only know how to look for them.

I've used TRIZ to understand problems and find solutions in the biosciences, local government, education, marketing, fast-moving consumer goods, industrial chemistry, engineering, software, management, construction, product design, packaging and more. The proof of the pudding's in the eating, so if you think TRIZ hasn't been used in your industry and want to see if it will work, be an innovator and try it yourself!

Waiting for the 'Right' Problem

Don't wait for a problem to come along that seems to match the TRIZ process perfectly. I've heard people say they wanted to wait until they found a suitably complicated problem that would show off TRIZ in the best way or they haven't yet got a problem hard enough to justify its use. TRIZ is a problem-solving toolkit – it can help you with any kind of problem.

Identify a problem or situation you're facing right now. Even if the issue seems trivial or you believe it can't be resolved with an ideal solution, as a result of lack of knowledge or implementation constraints, still give TRIZ a go. If applying new thinking to this problem will help you get started with the TRIZ tools, and you can learn from your experience, it will be time well spent. Don't worry about using TRIZ perfectly; simply use the tools, reflect on the process and the outcomes and then work out what you can do better next time. Start with the tools you feel most comfortable with and are excited by and become comfortable with using those first. You're better off using a bit of TRIZ a lot of the time than all the TRIZ tools never!

Starting Too Big

Trying to solve the most difficult, intractable problem that your industry's been grappling with for 30 years can be discouraging. While you're gaining familiarity and confidence with the TRIZ tools and processes, start with smaller problems, preferably urgent ones that you need to be looking at soon anyway. TRIZ is uniquely powerful for the most challenging problems, and it is very tempting to start there, but large, complex problems can be daunting when you're at the beginning of your TRIZ journey.

Work your way up to these very difficult problems, starting with problems that are a bit more manageable. If you had just taken up running, you wouldn't start by trying to run a marathon on your first day: you would train and maybe start with a few shorter races and work your way towards a marathon once you had developed your running skills and stamina. It's the same with problem solving: get familiar with the tools and the processes before you start on the hardest problem in your industry. That way, when you're working on the most difficult challenges, you can devote all of your considerable brain power to the problem at hand, not working out how to use the tools.

If you really want to start with something difficult, work on this difficult problem with someone experienced with TRIZ, so that person can guide you in the use of the tools where necessary.

Tackling Problems for Which You Can't Implement Solutions

World peace is desired by everybody, and the idea of applying TRIZ to make it a reality is appealing. Resolving conflict is a huge problem that nobody knows how to tackle – perhaps TRIZ could help? Applying TRIZ tools would give you greater understanding of the underlying problems that prevent world peace and also reveal some useful solutions. However, for most people, many of the solutions that TRIZ may come up with, such as changing diplomatic tactics, are beyond the scope of their control and problem solving becomes merely an academic exercise. (If any major world leaders are reading this book, however, do feel free to get in touch – I'm sure TRIZ can help!)

Looking at problems outside of your scope of influence – for example, changing the behaviour of your boss – will make you feel discouraged; you may devise genius solutions but you'll be unable to put them into practice. If you can reframe the problem to make it implementable by you, such as what you can do to help bring about less conflict or deal better with senior management, then it's worth tackling.

Tackling Problems without Involving the Problem Owner

If a problem needs to be solved, you need to involve the people who are responsible for that issue or area.

I've heard people suggest not involving problem owners because they were involved in its creation in the first place or they may be resistant to new ways of thinking about its resolution. In fact, that's why the problem owners *must* be involved. If they were on the ground as the problem manifested itself, they'll have crucial, detailed, inside knowledge that no one else possesses. You can use that information to help you better understand the problem and how it may be resolved. If you're looking for new ideas or ways of thinking, the problem owners can take your suggestions and consider what may prevent them from working; these issues can thus be tackled one by one at an early stage. Such problem-finding behaviour will be really valuable and should be regarded in a positive light.

The problem owners are the people responsible for implementing the solutions! People rarely like being told to implement someone else's ideas; being cut out of their development can make them feel disenfranchised, disengaged and annoyed. In contrast, being involved in the development of the solutions means they'll feel motivated to implement them, which will make the process smoother.

Trying to Solve Problems When You Don't Understand the Issue or the Technology

TRIZ is best with facts and real-world, nitty-gritty problems. If you don't actually understand how something works, you won't be able to apply TRIZ. To be fair, you won't be able to do any real problem solving without TRIZ either. This is a universal truth: you wouldn't ask a surgeon to build a bridge or construction engineers to do a heart transplant.

This is also true of parts of the problem: if you don't understand how individual elements of your system work, you need to either get the experts in the room or find out yourself. TRIZ will supplement, extend and make best use of your knowledge – it is *not* a substitute.

Trying to Solve Problems When You Lack Crucial Knowledge

This issue is subtly different to not understanding the issue or technology, described in the preceding section. Sometimes you need to know the relevant data to effectively tackle a problem; for example, to improve safety you need to know where, why and when the dangers exist in your situation or system; to come up with an improvement to a product, you need to know exactly who's buying it so you can better meet their needs; to streamline a process, you need to know how long each individual step in it takes; and to cut costs, you need to know exactly how much each individual element of your system costs – and all the useful things it does.

Before you can carry out any useful problem solving, you have to source the specific data needed. Without those data you won't be problem solving, you'll be guessing. Do some research. If the data are difficult to find or get, working out how to get them could be a TRIZzy problem to solve.

A caveat to this is that TRIZ can also help you work out what you need to know more precisely. Sometimes you know you don't have all the information required but you haven't worked out exactly what it is you need to know. TRIZ can help you map what you do know and highlight and define the areas where you need to find out more – and give you structured questions to find the information you need very efficiently.

Working on a Problem That's Already Been Solved

Sometimes people want to reapply the TRIZ process to a problem to see if it arrives at the same solution. Their logic is that they know the solution is innovative, and if TRIZ comes up with it again, it proves that TRIZ works. However, if you already know 'the answer', which actually is just one solution, you will then be afflicted by psychological inertia.

Working on problems as an academic exercise is less effective than dealing with real problems for two reasons:

✔ **The answer has already been thought of, and constrains everyone's thinking.** Every direction TRIZ suggests will seem to point to this solution, because that's what people have in their minds. If people explicitly try not to think about it, it's even worse; if I ask you to think of anything *except* a blue penguin, what's the first thing that comes to mind? Exactly.

✔ **People lack motivation.** It doesn't really matter if you find a solution or not; however, real problem solving is exciting partly because the outcome's unknown. That sense of challenge can be a great motivator and is absent when a problem's already been solved.

These problems can be circumvented if a different team works on the same problem, of course. However, creating a different team is almost impossible if you're working on a real problem because you still need to involve the problem owner and relevant subject matter experts.

Undertaking TRIZ by Stealth

TRIZ is useful for problem solving both on your own and with other people. However, if you try to cunningly lead your colleagues through the TRIZ problem-solving process without telling them the reason, they'll become suspicious about why you're trying to direct their thinking. That approach will be self-defeating anyway because they'll probably resist your efforts.

If you want to explicitly step through a TRIZ exercise with your colleagues, be open with them about the logic of what you're doing and why. Start small: for example, with a 30–60-minute session on capturing the Ideal Outcome (Chapter 9), or by charting the context of your situation in Time and Scale (Chapter 8), or by directed brainstorming with the 40 Inventive Principles to resolve a contradiction you've identified (Chapter 3).

This isn't to say that TRIZ thinking can't shape how you talk to others about problems in a useful way. Encouraging others to look at things in a new light – by being open to different ideas, searching for analogies, identifying the concepts behind ideas and talking about Ideal solutions – can be helpful for everyone, and you can pass on some of the TRIZ philosophy without any direct TRIZ training or exposure.

Giving Up Too Soon

If you're not getting the kind of solutions you want, take a break and then have another look at your problem. Have you scoped it correctly? Do you need to hit the problem with even more TRIZ (definitely)? Perhaps you'll find more innovative solutions at a level higher. Have you followed the process or skipped a few steps? If you have, go back and fill in the gaps. Is your Ideal Outcome really Ideal or did you try to make it more pragmatic? Your Ideal should stretch you. Go back and rework it. TRIZ helps you find solutions that exactly match your needs – if you don't start off with all the right needs, your TRIZ solutions will reflect this – including the gaps.

If you've come up with a range of imperfect solutions, develop them further. Consider constructing an Ideality Plot first (see Chapter 11); Maybe your ideas aren't as bad as you think. Start with those ideas that give you most of what you want even if they have big problems – in fact, *especially* if they have big problems. These are contradictions, and you know how to solve them! Now you have a route forward to even better solutions.

I've never worked through the TRIZ process and run out of something to do. You can always do more, and the more TRIZ work you do, the better your solutions will be. Biographies of great inventors always describe a point at which something went wrong and they didn't give up. Edison, da Vinci and pals kept trying, and eventually found what they needed. You now have the tools and the processes to help you do that too. Keep going!

Chapter 16

Ten Tips for Getting Started with TRIZ

In This Chapter

▶ Getting the best results with TRIZ

▶ Understanding the most effective TRIZ methods

*I*n this chapter you'll find simple hints and tips to get you started using TRIZ effectively and making TRIZ innovation a new habit. TRIZ has enormous rigour and depth, and sometimes people can feel a bit intimidated when they first encounter it . . . they don't know where to start putting it into practice. What follows are some recommendations based on my experience as a teacher and facilitator. And, of course, as someone who learned TRIZ herself (many moons ago)!

Learn It

You need to become familiar with the TRIZ tools before you can use them. Get to know the 40 Inventive Principles, the Standard Solutions, Trends and so on and make an effort to find new examples of these TRIZ tools put into practice.

Every morning, get out the list of the 40 Inventive Principles, pick one and try to think of examples of it in action. Do the same with the other tools and become familiar with what they mean. You don't need to learn them all by heart, but you do need to understand how they work in order to use them in real problem-solving situations.

When you're familiar with the tools, you can keep moving fast and focus all your energy and attention on the problem at hand, rather than on working out what the tools mean.

Use It

Nothing's as good as putting TRIZ into practice! Having a go at tackling a real issue will increase your understanding more than any amount of theory.

TRIZ is only useful when it's used – and the more you use it, the better you get. If anyone in your organisation is already using TRIZ, ask to come along and participate the next time they run a session. She'll probably be happy to have another pair of eyes on her problem, and you'll learn a lot from the other problem solvers in the room.

Start Small

Unless a TRIZ expert is going to help you step through the process, it's best not to start with a huge, complex problem because, if you struggle, it won't be clear whether you're having difficulty with the problem or the process.

A smaller problem is an easier place to start when you're trying to learn the tools as well as find solutions.

If you have an urgent but very thorny problem and want to hit it with some TRIZ, start small with the tools, do the thinking in stages, set realistic expectations and don't worry about whether you're doing it perfectly. Also, don't try to use every tool, perhaps just two or three to help you generate new thinking and fresh ideas. As you gain confidence and expertise, you'll be able to step through the whole process on larger problems.

Attend a Workshop

Learning a new technique is easier from a teacher than a book (even this one!). An expert can explain the concepts and help you put them into practice there and then.

Lots of opportunities to learn TRIZ exist out there, both face-to-face and online, including free webinars. The online article for Part V at www.dummies.com/extras/triz points you in the right direction.

Think and Talk TRIZ

Get TRIZ thinking into your normal speaking patterns and communicate with others using the TRIZ approaches. I'm not advocating that you spout TRIZ terminology at people whenever the words 'problem' or 'idea' come up, however. Rather, I mean use these terms to introduce TRIZ thinking into the conversation. For example, ask people to consider their Ideal outcome; what they'd want in an ideal world.

Think about analogies (does the current problem remind you or others of something similar?); ask who else may have solved similar problems in the past (maybe someone in your organisation has done so, for example); identify the concepts behind other people's ideas and suggest different ways of putting them into practice; ask whether reversing a process would solve the problem (Inventive Principle 13 – always a good place to start!); take things to extremes to get clearer thinking (applying the Size–Time–Cost creativity tool); consider how you could use your resources to get what you want (starting with the causes of the problem). (Inventive Principles are the subject of Chapter 3 and Chapter 7 covers the Size–Time–Cost tool.)

Changing how you talk about problems changes how you think about them, and the TRIZ philosophy will become second nature.

Find a Friend

Getting started with TRIZ can be difficult on your own. Find out whether anyone in your organisation, or even nearby geographically, is familiar with TRIZ and get together with her to discuss ideas or anything you're unsure of. If TRIZ sessions are already happening, people are usually more than happy to have an extra brain and additional pair of hands to help out. Observing others using the tools helps you translate theory – from this book or any other – into practice.

TRIZ conferences, where you can meet and share ideas with other TRIZniks, take place all over the world.

Finally, having a friend or colleague you trust join you when you're first trying to put the tools into practice can be helpful, as you can talk through how the tools work and the solutions you generate with someone else.

Fail Safely

The first time you run a TRIZ session at work, start small, with a friendly crowd, so that you feel 'safe to fail'. You'll learn a lot from a practice run and discover the best way to set up a session, describe the tools and explain what you want people to do.

Ask a colleague to be your helper and brief her thoroughly in advance on what you hope to get out of the session and what you want the participants to do. That way, she can help if people ask difficult questions or go off in unexpected directions.

If possible, send people some information about what TRIZ is in advance, and a short description of what you'll be doing and why, so if they have any questions about the process, you can answer them in advance and you don't find yourself answering tricky questions on your feet. Many people are willing to try something new, but if someone is really resistant, the invitation will give them a hint of what they'd be signing up for and they can excuse themselves in advance.

Be Bold

Applying TRIZ for the first time requires you to step out of your comfort zone and think in a different way, which may feel uncomfortable. TRIZ will suggest taking your problem solving in directions that may not immediately seem logical or practical or useful – but you still need to follow them.

You may discover a contradiction, for example, but the Inventive Principles (Chapter 3 covers the 40 Inventive Principles in detail) suggested for resolving it seem to bear no relation to your situation. Don't discard them! Doing what those Principles suggest means you'll generate the kind of solutions you wouldn't normally have thought of. This is exactly why you're using TRIZ: to generate new thinking.

When you've got a bit of experience under your belt, have a go at tackling a really hard, complex problem because these benefit the most from the application of TRIZ. Easier problems you can probably tackle on your own; it's for those where you get completely stuck or just can't see where you're going to find a solution that TRIZ will deliver the most value.

Fail Better

If you never fail, you're not being innovative enough!

Real innovation requires risk, and innovative organisations accept and forgive occasional failure as an inevitable by-product of trying new things. Failure can be harder to accept in yourself, however! But you learn most from experiences in which you fail, and if you're always trying to avoid mistakes, you'll play it safe and keep doing the things you've always done.

Applying TRIZ *should* take you out of your comfort zone, and part of the joy of using the TRIZ process is that it allows you to systematically explore uncertain and ambiguous situations, where innovative solutions are often found.

Many stories involve overcoming early failure to create great innovations. If something goes wrong when you're trying to put a solution into practice, don't give up! It's just another problem – and you can use TRIZ to solve it.

Reflect

After each TRIZ problem-solving session or project, give yourself time to reflect on what you did, what you learned and what you can do better next time.

It's okay if you haven't applied the tools perfectly so long as you can learn from the experience and use it as an opportunity to improve.

I always write down my 'lessons learned' from every TRIZ workshop I run; I recommend that you do the same, while the details of the session are still fresh in your mind, noting what went well or needs to be improved upon, unexpected questions that you need to find answers to and any useful information that you can use in your next TRIZ event.

Part VI
Appendixes

Visit www.dummies.com/cheatsheet/triz for a handy list of TRIZ basics that you can refer to anywhere you have Internet access.

In this part . . .

- The 40 Inventive Principles
- The Contradiction Matrix
- The 39 Parameters of the Contradiction Matrix
- The Separation Principles
- The Oxford TRIZ Standard Solutions

Appendix A
The 40 Inventive Principles

*T*he 40 Inventive Principles are the clever ways the world has found (so far) to solve contradictions. For example, one kind of contradiction is that you want something big and small – you might want something small at one time, but then big at another time, such as a ship in a bottle which has to be small to get into the bottle but then big when it is inside. This is typically done using Inventive Principle No.15, Dynamism (the boat folds down small to get through the narrow bottle neck, and then can be unfolded when it is inside the bottle).

Another example is an umbrella, which you want big when it is raining and small at all other times. Contradictions have always existed, and people have always found clever solutions to them. However, what TRIZ has done has catalogued *all* the ways of solving contradictions and distilled them down into a list of simple but clever concepts – the 40 Inventive Principles! If you want to read more about contradictions and the 40 Inventive Principles, check out Chapter 3.

Each of the 40 Inventive Principles has a number of *flavours* or suggestions; for example, Inventive Principle No. 1 suggests three ways your system can be segmented. This appendix includes all the Inventive Principles with technical, management and general examples of these principles put into practice.

Inventive Principle 1: Segmentation

✔ Divide an object into independent parts

- Flat-pack furniture
- Segmented tent poles
- Sales teams have different markets/geographical areas

✔ Make an object easy to take apart

- Quick-release bicycle wheels
- Use contract workers instead of permanent employees

✔ Make the object even more fragmented or segmented

- Many small, local offices rather than one large HQ

- Telecommute: employees work from home

Inventive Principle 2: Taking Out

✔ Take out a problem

- Air conditioning in the room, noise elsewhere

- Light pipes: get light where needed without heat

✔ Only get what you need

- Scarecrow

- Buy-in expertise

- No-frills hotel

Inventive Principle 3: Local Quality

✔ Change an object from uniform to non-uniform

- Ergonomic knife handles

- Offer different products or use different marketing strategies for different geographical markets

✔ Change the environment from uniform to non-uniform.

- Working hours phased to accommodate people working on international, shifted time-zone projects

✔ Make each part of an object function in conditions most suitable for its operation

- Night-time adjustment on rear-view mirror

✔ Make each part of an object fulfil a different and useful function

- Hammer with nail-puller

Inventive Principle 4: Asymmetry

✔ Change the shape of an object from symmetrical to asymmetrical

- Funnel with off-set exit for faster flow rate
- Use a different marketing approach for different types of clients

✔ Change the shape of an object to suit external asymmetries

- Ergonomic running shoes

✔ If an object is asymmetrical, make it more asymmetrical

- Different running shoes to suit users' running styles

Inventive Principle 5: Merging

✔ Join objects or operations together

- Bi-focal lenses
- Crew members in McDonalds

✔ Join parallel operations in time

- Include your customers and suppliers in design phase
- Work on a project in parallel rather than in series

Inventive Principle 6: Multi-Function

✔ Make an object perform multiple functions; eliminate the need for other objects.

- Swiss Army Knife
- 'One-stop shopping' – supermarkets sell insurance, banking services, fuel, newspapers and so on.

Inventive Principle 7: Nested Doll

✔ One object is placed inside another; which, in turn, is inside another, and so on.

- Shopping centres/malls

- Nested tables

- Measuring cups or spoons

✔ One object passes through another

- Telescopic car aerial

- Tape measure

Inventive Principle 8: Counterweight

✔ Compensate for the weight of an object by joining it with another object which provides lift.

- Hot-air balloons

- Kayaks with integrated foam floats

- Build teams of different personality types

✔ Compensate for the weight of an object by making it interact with the environment

- Maglev trains use magnetic levitation to create lift and propulsion, reducing friction and allowing for higher speeds

- Use renewable energy to reduce a company's carbon footprint

- Boost popularity by connecting with popular causes; for example, charities

Inventive Principle 9: Prior Counteraction

✔ Add a counteraction to manage a downside

- Make clay pigeons out of ice or clay

- Off-set carbon emissions

- Give generous severance packages when making redundancies

✔ Create beforehand stresses in an object that will oppose known undesirable working stresses later on

- Pre-stressed concrete compensates for concrete's weakness in tension
- Ergonomically assess workstations

Inventive Principle 10: Prior Action

✔ Do a required action in advance

- Practise emergency procedures in advance of crisis
- Agree meeting agendas in advance
- Arrange objects conveniently so they can go into action without loss of time
- Just-In-Time production
- In-store bakeries

Inventive Principle 11: Cushion in Advance

✔ Prepare emergency means beforehand to compensate for the relatively low reliability of an object

- Back up computer data
- Contingency planning
- Have more than one person trained in skills critical to the company

Inventive Principle 12: Equal Potential

✔ Change the conditions of work so that an object doesn't need to be raised or lowered

- Mechanic's pit in garage (car isn't lifted)
- Grow the job rather than promote the person

Inventive Principle 13: The Other Way Round

- ✔ Do the opposite action

    - Cash-back in stores: customers take money away instead of security firms

- ✔ Make movable parts fixed, and fixed parts movable

    - Home-shopping

    - Treadmills

    - Wind tunnels

- ✔ Turn the object upside down

    - Customers create their own product, for example, radio listeners dialling in for talk shows or to give traffic updates

    - Tomato sauce bottle with opening on the bottom

Inventive Principle 14: Spheres and Curves

- ✔ Switch from flat surfaces to spherical ones; from parts shaped as cubes or rectangles to ball-shaped structures

    - Mezzaluna (knife shaped as a half-moon)

    - Arches and domes for strength in architecture

    - 360-degree appraisals

- ✔ Use rollers, balls, spirals, domes

    - Rotary pizza cutter

    - Ballpoint pens

    - Archimedes screw

    - Make repeat purchases easy (such as direct debits, subscriptions)

- ✔ Go from linear to rotary motion, use centrifugal forces

    - Push–pull to rotary switches (for example, lighting dimmer switches)

    - Loyalty schemes

Inventive Principle 15: Dynamism

✔ Change the object or environment to work the best at every stage of work

- Adjustable steering wheel (or seat, back support, mirror position and so on)
- Different price and positioning for products throughout their life

✔ Divide an object into parts that can move relative to each other

- Bifurcated bicycle saddles
- Folding chairs

✔ If an object is rigid, make it movable

- Bendy drinking straw
- Hot-desking
- Virtual 360-degree tours online

Inventive Principle 16: Partial or Excessive Action

✔ If it's difficult to get 100 per cent of an action, go for more or less

- Overfill bottles on production line
- Overspray when painting, then remove the excess
- Ensure happy customers by providing exceptional customer service
- Show products and services online even if it's not possible to purchase online

Inventive Principle 17: Another Dimension

✔ Move to another dimension: from one to two or three dimensions

- Coiled telephone wire/garden hose

✔ Go from a single layer to multi-layers

- Multi-storey car park
- Hierarchy of command

✔ Tilt an object or turn it on its side

- Cars on road transporter inclined to save space

✔ Use the other side of an object

- Print on both sides of paper

- Electronic components mounted on both sides of circuit board

Inventive Principle 18: Mechanical Vibration

✔ Make an object vibrate

- Electric toothbrushes

✔ Increase the frequency of vibration (up to ultrasonic)

- Sonic toothbrushes

- Sonic facial brushes

✔ Use an object's resonant frequency

- Break up kidney stones with ultrasound

- Tuning fork

- Use piezoelectric vibrators instead of mechanical ones

- Quartz watches

✔ Use combined ultrasonic and electromagnetic field oscillations together

- Ultrasonic and electromagnetic pest repellers deter both mice and insects

Inventive Principle 19: Periodic Action

✔ Go from continuous action to periodic or pulsating actions

- ABS brakes

- Flashing bicycle lights

- Change leadership: many university heads of department only lead for a year, senior academics rotate leadership

✔ If an action is already periodic, change the rate or level of change

- • Appraise performance more regularly than annual reviews

✔ Use pauses between actions to perform a different action

- • Perform maintenance work during slow periods
- • Inkjet printer cleans head between passes

Inventive Principle 20: Continuous Useful Action

✔ Carry out work without a break

- • 24-hour manufacturing
- • Hospital emergency departments

✔ Remove idle or intermittent actions

- • Kayaks use double-ended paddle to utilise recovery stroke
- • Multi-skill workforce to enable them to perform other tasks when their core job has quiet periods

Inventive Principle 21: Rushing Through

✔ Do a process, or certain stages (which are harmful or dangerous) at high speed

- • Cut plastic faster than heat can propagate in the material, to avoid deforming the shape
- • Immediate dismissal

Inventive Principle 22: Blessing in Disguise

✔ Use harmful factors (particularly, harmful effects of the environment) to get something good

- • Pain can be useful feedback to stop doing something; for example, walk on broken ankle

✔ Remove a harmful action by adding another harmful action

- Reduce traffic in cities by introducing a congestion charge

✔ Increase a harmful factor so much that it is no longer harmful

- Restrict supply of goods to create scarcity value (such as designer handbags)

Inventive Principle 23: Feedback

✔ Introduce feedback

- Customer surveys

- Feedback forms after training

- Exercise apps that inform you of distance run, average speed, calories burned

✔ If feedback is already used, make it more effective

- Go from paper feedback forms to interviews or online surveys

- Share information from exercise apps on social media

Inventive Principle 24: Intermediary

✔ Use an intermediary object to pass on an action

- Play guitar with a plectrum

- Marriage counsellor

- Travel agent

✔ Temporarily join an object with another (easily removable)

- Oven gloves for hot dishes

- Bridging loans

- 'The A-Team'

Inventive Principle 25: Self-service

✔ An object services, maintains and repairs itself

- Self-cleaning ovens

- Wikipedia

- Customer loyalty reward schemes (such as Nectar cards) collect information about individual's purchasing decisions to target services and products

- Bicycle tyres filled with gel to seal punctures instantly

✔ Use waste (or lost) resources, energy or substances.

- Use heat from a process to generate electricity (co-generation)

- Use travel time to work

Inventive Principle 26: Copying

✔ Replace unavailable, expensive, complicated, fragile objects with cheap, simple copies

- Scan rare, historic books and documents, so they are accessible to all and the original remains protected

✔ Replace an object, or process with optical copies

- Measure an object from a photograph

- Product manuals as PDFs rather than printed booklets

- Online training instead of classroom training

✔ If visible optical copies are used, move to infrared or ultraviolet; use unusual ways of seeing/viewing situations

- Police helicopters use infrared to track suspects

- Evaluate customer satisfaction in multiple ways, for example, interviews, questionnaires, observing customers use product

Inventive Principle 27: Cheap, Short-Living Objects

✔ Replace an expensive object with many inexpensive objects that don't last as long

- Disposable serviettes, napkins, cups, cameras

- Automate work procedures and have low-skilled, low-paid staff who are easily replaceable

- Make product cheap and easily replaceable instead of reusable, for example, contact lenses, nappies, low-cost clothes

Inventive Principle 28: Replace Mechanical System

✔ Replace a mechanical system with a sensory one (optical, acoustic, taste or smell)

- Lights and bells rather than secure barriers at rail crossings

- Smell of baking bread to entice shoppers

- Security systems

✔ Use electric, magnetic and electromagnetic fields to interact with the object; use influence instead of rules

- Magnetic bearings

- TV remote control

- Use staff loyalty to encourage good behaviour

✔ Replace stationary fields with moving unstructured fields; replace unstructured fields with structured ones

- Hot desking

- GPS sensors inform central control point of location of delivery vans, taxis

✔ Use fields in conjunction with field-activated (such as ferromagnetic) particles

- Ultrasonic welding

Inventive Principle 29: Pneumatics and Hydraulics

✔ Use gas and liquid parts of an object instead of solid parts (for example, inflatable, filled with liquids, air cushion, hydrostatic, hydro-reactive); use flexible influences instead of solid rules

- Inflatable mattresses
- Hovercraft
- Use guidelines, not rules
- Create a 'health and safety culture' instead of a long list of rules

Inventive Principle 30: Flexible Membranes and Thin Films

✔ Use flexible shells and thin films instead of three-dimensional structures

- Tarpaulin car cover instead of garage
- Open-plan offices

✔ Isolate the object from the external environment using flexible shells and thin films

- Bubble-wrap
- Bandages and plasters
- Offices with cubicles

Inventive Principle 31: Porous Materials

✔ Make an object porous or add porous elements (inserts, coatings and so on)

- Cavity wall insulation
- Foam metals
- 'Open door' management policies

✔ If an object is already porous, use the pores to introduce a useful substance or function

- Medicated dressings

- Train customer service teams to sell additional products and services (for example, at shop tills)

Inventive Principle 32: Colour Change

✔ Change the colour of an object or its surroundings

- Colour-changing paint or sun cream

- Light-sensitive glasses

✔ Change the transparency of an object or its surroundings

- Make organisation's objectives and decision-making processes clear to everyone

- Transparent solar cells (make every window/screen a photovoltaic solar cell)

✔ Observe objects or processes that are hard to see by using coloured additives

- Use opposing colours to increase visibility (for example, butchers use green decoration to make meat look redder)

✔ If additives are already used, monitor things that are hard to see by adding luminescence/other markers

- UV marker pens to identify stolen goods (only seen under ultraviolet light)

- Get potential customers to register interest with special offers

Inventive Principle 33: Uniform Material

✔ Objects interacting with the main object should be made of the same material (or one with similar properties)

- Ice cubes made of drink they are cooling (such as lemonade)

- Wood dowel joints for joining wooden components

- Make sure all your employees can understand and sell your products

- 'Mirror' someone's body language to facilitate easy communication

Inventive Principle 34: Discarding and Recovering

✔ Make objects (or part of them) that have fulfilled their useful functions go away (discard by dissolving, evaporating and so on) or modify them directly during operation

- Dissolving capsules for vitamins and medicine

- Bio-degradable containers and bags

- Take on temporary staff to manage busy periods, such as in shops over Christmas

✔ Restore consumable parts of an object during operation

- Sell on to 'used up' customer: for example, transfer student bank accounts to graduate bank accounts when they finish their course; dentists refer patients to dental hygienists

- Self-sharpening blades

- Mechanical pencil

Inventive Principle 35: Parameter Change

✔ Change the physical state (for example, to a gas, liquid or solid)

- Transport gases as liquids

- Liquid soap instead of soap bar

- Deodorant as a liquid, solid stick or aerosol spray

- Virtual organisations

- Customer service offered remotely (by phone or chat box online) instead of face to face

✔ Change the concentration or density

- Change number of staff

- Fire briquettes: low density for lighting fires, high density for burning all night

- Move from local to centralised distribution (or vice versa)

✔ Change the degree of flexibility

- Vulcanised rubber is less flexible and stronger

- Hot-desking

- Flexi-time

✔ Change the temperature or volume

- 'Fire up' and motivate employees

- Increase individual's scope of responsibility

- Heated butter knives and ice cream scoops

- Butter knives with built-in grater to soften cold butter

✔ Change the pressure

- Pressure cooking is faster

- Manipulate stress levels (increase near deadlines)

✔ Change other parameters

- Semco: staff set their own wages, production targets, working hours

- Thixotropic paints are viscous so they don't drip, but become runny when shear force is applied by the brush against the surface being painted.

Inventive Principle 36: Phase Changes

✔ Use phenomena occurring during phase changes (volume changes, loss of absorption of heat and so on)

- Latent heat effects in melting or boiling

- Break rocks by soaking in water and then freezing

- Individuals try harder when proving themselves; for example, graduates, new to the post, newly promoted: use these people for difficult projects/those requiring long hours

Inventive Principle 37: Thermal Expansion

✔ Use expansion (or contraction) of materials by heat; responsiveness to circumstances

- Be very responsive to change; for example, have extra staff available for busy periods
- Emergency services available to deal with crises
- Shape memory materials
- Shrink-wrapping

✔ Use multiple materials with different coefficients of thermal expansion; use multiple and different systems that respond to circumstances differently

- Bi-metallic strips used in thermostats
- Different emergency services offer different expertise in major car crashes, for example, police create safe space on road and prevent other crashes; fire service to cut people out of cars; ambulance knows how to move injured people.

Inventive Principle 38: Boosted Interactions

✔ Enrich the atmosphere

- Create a competitive atmosphere to motivate sales team
- Oxygen tent for asthmatic patients
- Nitrous oxide injection for power boost in engines
- Create the right atmosphere for different working environments: places with buzz for team working and discussing ideas; quiet zones for independent working

✔ Create a highly enriched atmosphere

- Use highly charismatic leaders to engage the workforce
- Irradiation of food to extend shelf life

🖙 Enrich the atmosphere with unstable elements

- Staff charged with energy through uncertainty: such as fear (threat of redundancy) or rewards (bonuses/promotion)
- Use ozone to destroy micro-organisms

Inventive Principle 39: Inert Atmosphere

🖙 Replace a normal environment with an inert one

- Libraries' quiet environment creates a good environment to work, read and study
- Corporate away days
- Foam to separate a fire from oxygen

🖙 Add neutral parts or inert additives to an object

- Use contractors and external consultants
- Fire retardant additives

Inventive Principle 40: Composite Structures

🖙 Change from uniform to layered/composite (multiple) structures

- Teams with diverse team members bring different skills and perspectives
- Use different delivery methods in training (lectures, exercises, follow-up reading)
- Fibre-reinforced composite materials in Boeing 787 wing and fuselage
- Concrete aggregate

Appendix B

The Contradiction Matrix

*T*he TRIZ community identified that one kind of Contradiction is a Technical Contradiction: when you improve one thing, something else gets worse as a consequence. A huge amount of patent research uncovered that these Technical Contradictions and their solutions follow patterns. The Contradiction Matrix was developed from this analysis, so when you have a Technical Contradiction, you can look up in the Matrix which of the 40 Inventive Principles have been used most often in the past to create a clever solution. See Chapter 3 for more on how to uncover and define your Contradictions.

39 Technical Parameters

Improve this one without making this one worse

#	Parameter
1	Weight of moving object
2	Weight of stationary object
3	Length of moving object
4	Length of Stationary object
5	Area of moving object
6	Area of stationary object
7	Volume of moving object
8	Volume of stationary object
9	Speed
10	Force (intensity)
11	Stress or pressure
12	Shape
13	Stability of the object's composition
14	Strength
15	Duration of action of a moving object
16	Duration of action of a stationary object
17	Temperature
18	Illumination Intensity
19	Use of energy by a moving object
20	Use of energy by a stationary object
21	Power
22	Loss of energy
23	Loss of substance
24	Loss of information
25	Loss of time
26	Quantity of substance
27	Reliability
28	Measurement accuracy
29	Manufacturing precision
30	Object-affected harmful factors
31	Object-generated harmful factors
32	Ease of manufacture
33	Convenience of Use
34	Ease of repair
35	Adaptability or versatility
36	Device complexity
37	Difficulty of detecting and measuring
38	Extent of automation
39	Productivity

Appendix C

The 39 Parameters of the Contradiction Matrix

· ·

*T*he Contradiction Matrix and the 40 Inventive Principles are helpful for solving contradictions. The Matrix consists of 39 improving and worsening parameters.

The following definitions are based on the *TRIZ Journal* (21 November 1998's entry; see www.triz-journal.com) by Ellen Domb, from translations of Altshuller's work:

✔ **Moving object:** Objects that can easily change position in space, either on their own, or as a result of external forces. Examples are vehicles, objects designed to be portable and objects which are used while in motion.

✔ **Stationary object:** Objects that do not change position in space, either on their own or as a result of external forces.

If you're not sure whether your system is movable or stationary (for example, a wind turbine), consider whether the part of the problem you are looking at is moving while it is in use. If, for example, you are looking at the wind turbine as a whole, it is stationary; if you are interested in a problem with the blades, they are moving. If you're not sure, try both!

Number	Title	Explanation
1	Weight of moving object	The mass of the object, in a gravitational field. The force that the body exerts on its support or suspension.
2	Weight of stationary object	The mass of the object, in a gravitational field. The force that the body exerts on its support or suspension, or on the surface on which it rests.

(continued)

Number	Title	Explanation
3	Length of moving object	Any one linear dimension, not necessarily the longest, is considered a length.
4	Length of stationary object	Same as above.
5	Area of moving object	A geometrical characteristic described by the part of a plane enclosed by a line, the part of a surface occupied by the object or the square measure of the surface, either internal or external, of an object.
6	Area of stationary object	Same as above.
7	Volume of moving object	The cubic measure of space occupied by the object. Length × width × height for a rectangular object, height × area for a cylinder and so on.
8	Volume of stationary object	Same as above.
9	Speed	The velocity of an object; the rate of a process or action in time.
10	Force	Force measures the interaction between systems. In Newtonian physics, force = mass × acceleration. In TRIZ, force is any interaction that is intended to change an object's condition.
11	Stress or pressure	Force per unit area. Also, tension.
12	Shape	The external contours, appearance of a system.
13	Stability of the object's composition	The wholeness or integrity of the system; the relationship of the system's constituent elements. Wear, chemical decomposition and disassembly are all decreases in stability. Increasing entropy is decreasing stability.
14	Strength	The extent to which the object is able to resist changing in response to force. Resistance to breaking.
15	Duration of action by a moving object	The time that the object can perform the action. Service life. Mean time between failure is a measure of the duration of action. Also, durability.
16	Duration of action by a stationary object	Same as above

Number	Title	Explanation
17	Temperature	The thermal condition of the object or system. Loosely includes other thermal parameters, such as heat capacity, that affect the rate of change of temperature.
18	Illumination intensity	Light flux per unit area, also any other illumination characteristics of the system such as brightness, light quality and so on.
19	Use of energy by moving object	The measure of the object's capacity for doing work. In classical mechanics, energy is the product of force × distance. This includes the use of energy provided by the super-system (such as electrical energy or heat). Energy required to do a particular job.
20	Use of energy by stationary object	Same as above.
21	Power	The time rate at which work is performed. The rate of use of energy.
22	Loss of energy	Use of energy that does not contribute to the job being done. See 19. Reducing the loss of energy sometimes requires different techniques from improving the use of energy, which is why this is a separate category.
23	Loss of substance	Partial or complete, permanent or temporary, loss of some of a system's materials, substances, parts or subsystems.
24	Loss of information	Partial or complete, permanent or temporary, loss of data or access to data in or by a system. Frequently includes sensory data such as aroma, texture and so on.
25	Loss of time	Time is the duration of an activity. Improving the loss of time means reducing the time taken for the activity. 'Cycle time reduction' is a common term.
26	Quantity of substance/the matter	The number or amount of a system's materials, substances, parts or subsystems that might be changed fully or partially, permanently or temporarily.
27	Reliability	A system's ability to perform its intended functions in predictable ways and conditions.

(continued)

Number	Title	Explanation
28	Measurement accuracy	The closeness of the measured value to the actual value of a property of a system. Reducing the error in a measurement increases the accuracy of the measurement.
29	Manufacturing precision	The extent to which the actual characteristics of the system or object match the specified or required characteristics.
30	External harm affects the object	Susceptibility of a system to externally generated (harmful) effects.
31	Object-generated harmful factors	A harmful effect is one that reduces the efficiency or quality of the functioning of the object or system. These harmful effects are generated by the object or system as part of its operation.
32	Ease of manufacture	The degree of facility, comfort or effortlessness in manufacturing or fabricating the object/system.
33	Ease of operation	Simplicity: The process is not easy if it requires a large number of people, large number of steps in the operation, needs special tools and so on. 'Hard' processes have low yield and 'easy' processes have high yield; they are easy to do right.
34	Ease of repair	Quality characteristics such as convenience, comfort, simplicity and time to repair faults, failures or defects in a system.
35	Adaptability or versatility	The extent to which a system/object positively responds to external changes. Also, a system that can be used in multiple ways under a variety of circumstances.
36	Device complexity	The number and diversity of elements and element interrelationships within a system. The user may be an element of the system that increases the complexity. The difficulty of mastering the system is a measure of its complexity.
37	Difficulty of detecting and measuring	Measuring or monitoring systems that are complex, costly, require much time and labour to set up and use, or that have complex relationships between components or components that interfere with each other all demonstrate 'difficulty of detecting and measuring.' Increasing cost of measuring to a satisfactory error is also a sign of increased difficulty of measuring.

Number	Title	Explanation
38	Extent of automation	The extent to which a system or object performs its functions without human interface. The lowest level of automation is the use of a manually operated tool. For intermediate levels, humans program the tool, observe its operation and interrupt or re-program as needed. For the highest level, the machine senses the operation needed, programs itself and monitors its own operations.
39	Productivity	The number of functions or operations performed by a system per unit time. The time for a unit function or operation. The output per unit time, or the cost per unit output.

Appendix D

The Separation Principles

*O*nce you have identified that you have a Physical Contradiction (see Chapter 3), you need to work out how to separate what you want in Time (T), Scale (S) or Condition (C). You can then look up in the table below which of the 40 Inventive Principles will be most useful for generating solutions.

Inventive Principle	Separate in Time	Separate in Space	Separate on Condition	Separate by System
1. Segmentation	T	S		S
2. Taking Out		S		
3. Local Quality		S		S
4. Asymmetry		S		
5. Merging				S
6. Multi-Function				S
7. Nested Doll	T	S		
8. Counterweight				S
9. Prior Counteraction	T			
10. Prior Action	T			
11. Cushion in Advance	T			
12. Equal Potential				S
13. The Other Way Round		S		S
14. Spheres and Curves		S		
15. Dynamism	T			
16. Partial or Excessive Action	T			
17. Another Dimension		S		

(continued)

Inventive Principle	Separate in Time	Separate in Space	Separate on Condition	Separate by System
18. Mechanical Vibration	T			
19. Periodic Action	T			
20. Continuous Useful Action	T			
21. Rushing Through	T			
22. Blessing in Disguise				S
23. Feedback				S
24. Intermediary	T	S		
25. Self-service				S
26. Copying	T	S		
27. Cheap, Short-Living Objects	T			S
28. Replace Mechanical System			C	
29. Pneumatics and Hydraulics	T		C	
30. Flexible Membranes and Thin Films		S		
31. Porous Materials			C	
32. Colour Change			C	
33. Uniform Material				S
34. Discarding and Recovering	T			
35. Parameter Change			C	
36. Phase Changes			C	
37. Thermal Expansion	T			
38. Boosted Interactions			C	
39. Inert Atmosphere			C	
40. Composite Materials		S		S

Appendix E

The Oxford TRIZ Standard Solutions

*T*he Oxford TRIZ Standard Solutions are a powerful tool for dealing with problems: they are simple lists of solutions based on analysis of patents and scientific journals. They are organised according to the kind of problem they solve:

- ✔ Dealing with harm (H)
- ✔ Overcoming insufficiency (i or i.a.)
- ✔ Difficulties with measuring and detecting (M)

The Standard Solutions are numbered in a logical way. Each Standard Solution has both one or two letters and two numbers: the letter tells you the class of the Standard Solution (such as H for 'harm'); the first number tells you the category of Standard Solution (for example, 2 for 'Block the harm'), the second number identifies the specific Standard Solution (such as 'protect part of the system from harm').

Dealing with Harm

H1 = Trim out the harm (6 solutions, so H1.1, H1.2 and so on until H1.6)

H2 = Block the harm (11 solutions)

H3 = Turn harm into good (4 solutions)

H4 = Correct the harm afterwards (3 solutions)

Overcoming Insufficiency

i1 = Add something to the subject or object (7 solutions)

i2 = Evolve the subject and object (10 solutions)

i.a. = Improve the action (18 solutions)

Difficulties with Measuring or Detecting

M1 = Indirect methods (3 solutions)

M2 = Add something (4 solutions)

M3 = Enhance measurement with fields (3 solutions)

M4 = Use additives with fields (5 solutions)

M5 = Evolve the measurement system (2 solutions)

Solutions for Dealing with Harms

Four strategies help you to deal with harmful actions, which can be achieved in 24 solutions:

✔ **Trim out the harm: 6 ways**

✔ **Block the harm: 11 ways**

✔ **Turn the harm into good: 4 ways**

✔ **Correct the harm afterwards: 3 ways**

H1 Trim out the harm

Remove components that have harms associated with them by understanding all the actions delivered by the component, both good and bad. Then transfer the useful actions to other parts of the system, allowing the component to be removed: creating a simpler system of fewer components that delivers all useful actions with fewer harms (and associated costs). Trimming is typically performed after a Function Analysis (Chapter 12) has been completed.

Start by trimming components that

✔ Are long way from the Prime Output

✔ Have lots of harmful actions

✔ Are expensive or unnecessarily complicated

Before transferring the useful actions, ask whether you need the object of the useful action. If the object is not required, you can trim both the subject and the object.

H1.1 Do we need the useful action of the component? If not, remove the component.

H1.2 Could the object perform the useful action? If yes, trim the subject and transfer responsibility for the useful action to the object.

H1.3 Could another component perform the useful action? If yes, trim the subject and transfer responsibility for the useful action to the other component.

H1.4 Could a resource perform the useful action? If yes, trim the subject and transfer responsibility for the useful action to the resource.

H1.5 Could we trim the subject after it has performed its useful action? If yes, trim the subject in time.

H1.6 Could we partially trim any harmful parts but leave any useful parts? If yes, trim in space: remove only the part of the subject that is delivering the harmful action, or only the part of the object that is receiving the harmful action.

H2 Block the harm

H2.1 Counteract the harmful action with an opposing field that neutralises the harm. For example, refrigerator keeps food cool in hot weather; noise-cancelling headphones work by producing a noise-cancelling wave that is 180° out of phase with the ambient noise.

H2.2 Change the object so it is not sensitive to the harmful actions. For example, add vents in an umbrella so wind doesn't make it turn inside out.

H2.3. Change the zone and/or duration of the harmful action to decrease its effects. For example, reduce the harmful effects of X-rays by giving the minimum effective dose required to obtain a clear image, in only the place that it is needed.

H2.4 Insulate from the harmful action by introducing a new component or substance; this can be made from elements of existing components that can be modified (including voids, bubbles, foam and so on). For example, rind on cheese; use surgical gloves to prevent infecting patients.

H2.5 Introduce a sacrificial substance to absorb the harm. For example, bumpers on cars; steel mesh balls in kettles to attract limescale and prevent build-up on the heating element.

H2.6 If a required action harms the system, apply it indirectly via a linked element. For example, oven gloves; melt chocolate in a bain marie (double boiler).

H2.7 Protect part of the system from harm; for example, lead aprons during X-rays.

H2.8 Reduce the harm by using a weaker action and enhancing it only where needed. For example, knowledge speed cameras may be present on roads reduces driver speeds generally: actual cameras are placed at accident black spots to reinforce the effect.

H2.9 Use sub-systems/details of components to stop the harm. For example, silver used in socks to kill odour-causing bacteria.

H2.10 Use super-systems/the environment to stop the harm. For example, air conditioning to cool rooms that contain a lot of computer equipment.

H2.11 Switch off the harm. For example, remove magnetism of steel tools; medication that doesn't allow absorption of fat from food 'switches off' the harm of fat.

H3 Turn the harm into good

H3.1. Use the harm to deal with the harm. For example, vaccination (small component of a virus stimulates immune response); deliberately burning a strip of land ahead of a wildfire.

H3.2 Use the harm for something good. For example, use waste food and grass cuttings for compost; stress of exams promotes hard work and learning material.

H3.3. Add another harm so that the combination of the two harms is no longer harmful. For example, add a strong alkali to a strong acid to neutralise.

H3.4 Amplify the harm until it delivers a benefit. For example, extinguish an oil well fire by creating an explosion at the well head. This momentarily consumes the atmospheric oxygen and puts out the fire; encourage extremist politicians to air their views on every topic to make their extremism very clear.

H4 Correct the harm

H4.1 After a harmful action has happened, eliminate any of its harmful consequences. For example, kitchen paper towels to mop up spills; after a traumatic event, seek counselling to mitigate long-term psychological impact.

H4.2 Counteract or control the harmful effects from a harmful action – either during or beforehand. For example, pre-stressed concrete; training for frontline emergency staff in dealing with difficult/violent people.

H4.3 Predict and eliminate harms before they happen, use something that disappears or becomes part of the system/environment. For example, use ice instead of sand for cleaning buildings; prevent potential damage from removing stitches by using dissolvable stitches in wounds.

Solutions for Improving Insufficiency

There are two strategies, delivered by 35 solutions:

> ✔ **Improve the components:**
>
> • Add something to the subject or the object: 7 ways
>
> • Evolve the components: 8 ways
>
> ✔ **Improve the action: 18 ways**

i1 Add something to the subject or object

i1.1 Add something to or inside the subject or object to enhance the function. The additive can be permanent or temporary (if temporary it disappears or decomposes after use). For example, metal skewers in baking potatoes to help heat reach the middle of the potato.

i1.2 Add something between the subject and object which enhances the function. For example, spectacles to enhance eyes; wrapping baking potato in foil to speed up cooking; shoe horn.

i1.3 Use the external environment to enhance/provide the functions. For example, fluoride in drinking water.

i1.4 Change or add something to the environment/surroundings of the subject and object. For example, add smoke to wind tunnel to monitor air flow around an object more effectively.

i1.5 Add something outside/around the subject and object to change the features/properties of the subject or object or provide extra functions such as protection. For example, sugar coating on chocolates to prevent melting.

i1.6 Add something from the environment/surroundings to enhance the function. For example, sea water used as ballast in ships.

i1.7 If we can't add anything, use the deterioration or decomposition of the components/environment to enhance the function. For example, green patina caused by oxidation protects copper roofs from further wear; use the two halves of a broken eggshell to help separate white from yolk.

i2 Evolve the subject and object

i2.1 Segment the subject or object; increase the degree of fragmentation/ divide into smaller units. For example, multiple cylinders in internal combustion engines; grind coffee beans.

i2.2. Introduce voids, fields, air, bubbles, foam or similar into the subject or object. For example, wetsuits trap water to provide insulation; bubble wrap; duvets.

i2.3 Improve systems by multiplying similar system elements, or combining with another similar system to improve actions; add new dissimilar elements to provide extra functions which will become integrated as the system evolves. For example, to TVs add speakers, DVD players, digital TV boxes, streaming capability from other devices; Swiss army knife; combined washing machine and dryer.

i2.4 Make the system more flexible/adaptable/dynamic. For example, business reorganisation from monolithic to smaller, focused business units; expanding via franchising, licensing and partnering on specific projects; cut-to-size insoles for shoes; clingfilm.

i2.5 Develop the links between the system elements (make them more flexible or more rigid). For example, snow chains on car tyres; Rubik cube.

i2.6 Increase the difference between elements. For example, seek out diverse employees (age, experience, background) to get a varied workforce with a wider spread of ideas.

i2.7 Separate by scale; provide opposite functions at different levels. For example, a bicycle chain has rigid components to transmit force but the system as a whole is flexible.

i2.8 Transition function delivery to the micro-level. For example, self-cleaning glass; microcapsules in cosmetics; microwave ovens.

i.2.9 Improve controllability by developing a component or part of the system to deliver additional useful functions. For example, central heating thermostat that incorporates a timer to give different set points at different times (such as cooler overnight).

i2.10 Improve a system by changing components/substances to deliver exactly what is needed in time and/or space. For example, Direct Line Feed: component suppliers deliver parts directly to a customer's production line.

i.a. Improve the action

These solutions help us get the right action and get the action right; they are useful when we need either to improve an existing action or to get an action that is currently missing.

i.a.1 Add a missing action; also add a subject if required. Look in the TRIZ Effects Database if required. For example, find a way of chilled food telling us it is still safe to eat (more precisely than use-by dates); add wings to F1 racing cars to provide downward force and increased cornering grip.

i.a.2 Add another action from existing components. For example, sonic toothbrushes use sound waves as well as mechanical vibration to clean teeth.

i.a.3 Change the action to a better one. For example, induction hobs use electromagnetic induction to heat pans rather than thermal conduction, providing more efficient and safer heating.

i.a.4 Change from a uniform action to an action with predetermined patterns. For example, pressure washer hoses pulse water to improve cleaning.

i.a.5 Match or mismatch the natural frequency of actions with the natural frequency of the subject or the object. For example, hot desk part-time employees.

i.a.6 Match or mismatch the frequencies of different actions. For example, a production line of different people doing different actions works much more efficiently than everyone working in parallel.

i.a.7 To achieve two incompatible actions, perform one action in the downtime of the other. For example, maintenance and cleaning of office buildings at night.

i.a.8 Use existing actions/fields to create other actions/fields. For example, use the heat from lightbulbs to heat a cold bathroom.

i.a.9 Use actions that exist in the environment (such as gravity, ambient temperature, pressure, sunlight). For example, solar-powered torches.

i.a.10 Get another action from any available resources such as other components in your system, or the environment. For example, use your smartphone to control your home heating.

i.a.11 Use excessive action and remove the surplus. For example, overfill pints of beer to ensure a creamy head of foam and remove the excess.

i.a.12 Use a small amount of a very active additive. For example, stain removing powders added to clothes washes.

i.a.13 Concentrate the additive at a specific location. For example, spot treatment of stains on clothing.

i.a.14 Introduce an additive temporarily. For example, flavour stews with herbs in a muslin bag, which can be removed.

i.a.15 Use a copy or mode of the object in which additives can be used, instead of the original object, if additives are not permitted in the original. For example, electronic cigarettes.

i.a.16 Improve an action by changing the phase of the existing action or component (use solid, liquid or gas of the same material). For example, steam cleaning; transport natural gas as a cryogenic liquid.

i.a.17 Achieve an action by using phenomena that accompany phase change. For example, split rocks by filling with water then reducing temperature below zero: they crack when the water turns to ice.

i.a.18 Achieve an action with dual properties by using components capable of converting from one phase state to another. For example, shape-memory metal heat exchangers.

Solutions for Detection and Measurement

The recommendations for improving measurement and detection are arranged as five subclasses, containing 17 Standard Solutions:

- ✔ **M1 Indirect Methods: 3 solutions**

- ✔ **M2 Add Something: 4 solutions**

- ✔ **M3 Enhance Measurement with Fields: 3 solutions**

- ✔ **M4 Use Additives with Fields: 5 solutions**

- ✔ **M5 Evolve the Measurement System: 2 solutions**

With each of the subclasses, you start with the first suggestion, as this will be the most radical. You move down to the next suggestion if it isn't possible to implement the first suggestion, until you find a solution. Even if you find solutions earlier, it will be worth seeing if the later suggestions will generate any additional ideas.

M1 Indirect methods

M1.1 Change the problem so there is no need for measurement or detection. For example, cooking in boiling water doesn't require measuring temperature: boiling water is always at 100°C.

M1.2 Measure a copy or image. For example, use photos or mirrors; ultrasound scans are used to measure the size of growing foetuses; barcodes and scanners to give product prices.

M1.3 Change the problem into detecting or measuring a number of consecutive, successive changes. For example, calculate when the Big Bang happened from the current distance and velocity of galaxies.

M2 Add something

M2.1 Add something and then measure changes to one of its features. For example, there are egg timers that you place in the same saucepan of water as your egg, which change colour to tell you when the egg is cooked; car tyre tread-wear indicators.

M2.2 Add something made from an extra/new component, which creates a field. For example, Mercaptan (a substance that smells strongly of rotten eggs) is added to natural gas (which is odourless) to enable leak detection.

M2.3 Add something to the environment around our system that reacts to what you want to measure. For example, put leaking bicycle tyre in water to find leak.

M2.4 Create additives in the environment by decomposing or changing the environment. For example, track location of aeroplanes from vapour trails.

M3 Enhance measurement with fields

M3.1 Use the natural phenomena that are already a part of our system; utilise the scientific/physical effects to observe changes. For example, measure human body temperature by measuring intensity of the emitted infrared radiation; when boiling sugar, observe the different characteristics of the sugar (soft-ball, hard-ball, hard-crack and so on) to know when to stop, instead of using a thermometer.

M3.2 Use resonance or resonant frequency. For example, observe changes in the resonant frequency of a system, component or the environment that are related to what you want to measure. For example, detect cracks in bells or bowls by striking them: the note sounds different when cracked.

M3.3 Join what you want to measure to another object and measure its resonant frequency, or measure the resonance in its environment. For example, running a wetted finger around the rim of a wine glass generates a note whose pitch is related to the amount of water in the glass.

M4 Use additives with fields

M4.1 Add or make use of a substance that has a measurable field. For example, substances with magnetic fields.

M4.2 Add easily detectable substance or particles with a measurable field to your system. For example, luminous paint.

M4.3 Add easily detectable substance or particles with a measurable field inside your system. For example, put ferromagnetic particles inside your system.

M4.4 Add easily detectable particles with a measurable field to the environment. For example, throw grass in the air to measure wind speed and direction.

M4.5 Use scientific/physical effects. For example, Curie point, Hopkins, Barkhausen.

M5 Evolve the measurement system

M5.1 Go from one measurement to two or more to improve the quality of relying on a single measurement. The individual measurements may be separated

- ✔ In time (for example, use multiple measurements to compensate for an unreliable or low-resolution sensor
- ✔ In space (for example, a widely spaced array of small telescopes can obtain or exceed the performance of a single large instrument)
- ✔ By measurement type (for example, multi-spectral analysis)

M5.2 Measure indirectly from the first and then second derivatives in time or space. For example, measure distance of stars from their brightness (first derivative) and colour (second derivative). The brightness of a star is dependent on the distance from us; the colour then changes as a result of the different brightness. By identifying the colour of a star, you can work out the brightness . . . and from that, the distance.

Index

About the Author

Lilly Haines-Gadd's life changed as a result of TRIZ. After a music degree at King's College London, she worked as a management consultant, solving business problems for telecoms companies. When she joined Oxford Creativity in 2004 it was to run the business, but she discovered TRIZ was useful for everyone – and through applying TRIZ on business problems, she found that her way of thinking had totally changed. This led to her retraining in Psychology, studying at Oxford Brookes, Cardiff University, and completing an MSc in Occupational Psychology at Birkbeck College, University of London. Her dissertation was evaluating the impact of TRIZ training at an individual and organisational level. She is trained in delivering various psychometric measures and is a member of the British Psychological Society.

In addition to running the company, she delivers training, facilitates innovative problem-solving workshops and develops new workshops, materials and methods for learning TRIZ. She is interested in how TRIZ makes everyone a better problem solver and a more creative thinker – whatever their natural approach to problems. Through her extensive hands-on experience of problem solving and TRIZ training, she has observed that TRIZ not only gives individuals great clarity of thought but also gets diverse teams working together, communicating more effectively and thinking differently.

Lilly is still a keen musician, singing in several choirs in Oxford, and is improving her Italian.

Dedication

To Ken Gadd.

Author's Acknowledgements

Firstly, to my mother, Karen Gadd, for introducing me to TRIZ and supporting and encouraging me throughout my TRIZ journey. It's been a real joy to work with her on developing new ways of teaching and using TRIZ. Without her, this book wouldn't have been possible.

My colleagues at Oxford Creativity have also contributed much to our work and logic: Andrew Martin, Neil Sherry, Andrea Mica, Ron Donaldson and Frederic Mathis, Merryn Haines-Gadd, Geoffrey Haines and Michael Haines have been enormously helpful in providing, discussing and refining ideas for the book.

There have been many others in the TRIZ community who have informed my thinking and taught me a great deal about TRIZ, most notably Dr Sergei Ikovenko and Ellen Domb.

The people I have taught and facilitated in TRIZ workshops have all taught me something new, but some of my clients and friends in particular have provided examples, helpful feedback and fresh thinking, including Neal Symmons, Peter Knowles, Kent Haell, Mark Veevers, David Drummond, Alastair Clarke, Mike McMenamin, Emily Lloyd and Rebecca Rue.

Enrico Sorrentino has been a constant source of inspiration and support.

Thanks to Annie Knight, Vicki Adang, Kate O'Leary, Daniel Mersey of Word Mountain Creative Content, Iona Everson, Rachael Chilvers and the rest of the team at Wiley for their unfailingly cheerful guidance and hard work throughout the process of preparing this book.

Publisher's Acknowledgements

Commissioning Editor: Annie Knight

Project Manager: Victoria M. Adang

Development Editor: Daniel Mersey of Word Mountain Creative Content

Copy Editor: Kate O'Leary

Technical Editor: Karen Gadd

Art Coordinator: Alicia B. South

Production Editor: Kumar Chellappan

Cover Image: © yod67/Shutterstock

Take Dummies with you everywhere you go!

Whether you're excited about e-books, want more from the web, must have your mobile apps, or swept up in social media, Dummies makes everything easier.

FOR DUMMIES®
A Wiley Brand

BUSINESS

978-1-118-73077-5

978-1-118-44349-1

978-1-119-97527-4

MUSIC

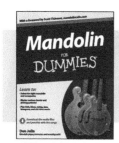

978-1-119-94276-4

978-0-470-97799-6

978-0-470-49644-2

DIGITAL PHOTOGRAPHY

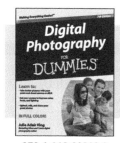

978-1-118-09203-3

978-0-470-76878-5

978-1-118-00472-2

Algebra I For Dummies
978-0-470-55964-2

Anatomy & Physiology For Dummies, 2nd Edition
978-0-470-92326-9

Asperger's Syndrome For Dummies
978-0-470-66087-4

Basic Maths For Dummies
978-1-119-97452-9

Body Language For Dummies, 2nd Edition
978-1-119-95351-7

Bookkeeping For Dummies, 3rd Edition
978-1-118-34689-1

British Sign Language For Dummies
978-0-470-69477-0

Cricket for Dummies, 2nd Edition
978-1-118-48032-8

Currency Trading For Dummies, 2nd Edition
978-1-118-01851-4

Cycling For Dummies
978-1-118-36435-2

Diabetes For Dummies, 3rd Edition
978-0-470-97711-8

eBay For Dummies, 3rd Edition
978-1-119-94122-4

Electronics For Dummies All-in-One For Dummies
978-1-118-58973-1

English Grammar For Dummies
978-0-470-05752-0

French For Dummies, 2nd Edition
978-1-118-00464-7

Guitar For Dummies, 3rd Edition
978-1-118-11554-1

IBS For Dummies
978-0-470-51737-6

Keeping Chickens For Dummies
978-1-119-99417-6

Knitting For Dummies, 3rd Edition
978-1-118-66151-2

FOR DUMMIES
A Wiley Brand

SELF-HELP

978-0-470-66541-1

978-1-119-99264-6

978-0-470-66086-7

LANGUAGES

978-0-470-68815-1

978-1-119-97959-3

978-0-470-69477-0

HISTORY

978-0-470-68792-5

978-0-470-74783-4

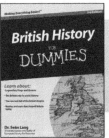

978-0-470-97819-1

Laptops For Dummies 5th Edition
978-1-118-11533-6

Management For Dummies, 2nd Edition
978-0-470-97769-9

Nutrition For Dummies, 2nd Edition
978-0-470-97276-2

Office 2013 For Dummies
978-1-118-49715-9

Organic Gardening For Dummies
978-1-119-97706-3

Origami Kit For Dummies
978-0-470-75857-1

Overcoming Depression For Dummies
978-0-470-69430-5

Physics I For Dummies
978-0-470-90324-7

Project Management For Dummies
978-0-470-71119-4

Psychology Statistics For Dummies
978-1-119-95287-9

Renting Out Your Property For Dummies, 3rd Edition
978-1-119-97640-0

Rugby Union For Dummies, 3rd Edition
978-1-119-99092-5

Stargazing For Dummies
978-1-118-41156-8

Teaching English as a Foreign Language For Dummies
978-0-470-74576-2

Time Management For Dummies
978-0-470-77765-7

Training Your Brain For Dummies
978-0-470-97449-0

Voice and Speaking Skills For Dummies
978-1-119-94512-3

Wedding Planning For Dummies
978-1-118-69951-5

WordPress For Dummies, 5th Edition
978-1-118-38318-6

Think you can't learn it in a day? Think again!

The *In a Day* e-book series from *For Dummies* gives you quick and easy access to learn a new skill, brush up on a hobby, or enhance your personal or professional life — all in a day. Easy!

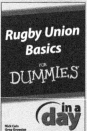

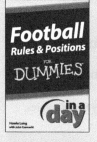

Printed and bound by CPI Group (UK) Ltd, Croydon, CR0 4YY

27/10/2024

14580395-0001